Running feet, sharp noses

Running feet, sharp noses

Essays on the animal world

PVA Books

First published in 2023 by PVA Books.

Copyright © 2023 individual contributors,
Paper Visual Art Journal

All rights reserved. No part of this publication may be reproduced in whole or in part in any form without prior written permission from the publisher.

'Opening a Gate' by John Berger was published in *The Shape of a Pocket* © 2001. Reproduced with permission of Pantheon Books, an imprint of the Knopf Doubleday Publishing Group, a division of Penguin Random House LLC, and of Bloomsbury Publishing PLC.

'The Courage of Turtles' by Edward Hoagland was originally published in the *Village Voice* in 1968 and then in his eponymously titled book (Penguin Random House, 1970). Reproduced with permission of the author.

An earlier version of 'The Roaming Edges' by Suzanne Walsh was published in *Response to a Request*, ed. Rebecca O'Dwyer, in 2017. Reproduced with permission.

ISBN 978-1-9161509-4-2

Designed by Daly & Lyon, London
Printed in Germany by GGP Media GmbH, Pößneck

PVA greatly acknowledges the support of the Arts Council / An Chomhairle Ealaíon.

papervisualart.com

For Brussels, Cara, Cloudy, Dougal, Eddie,
Eli, Feeby, Flash, Freddie, Ginger, George, Gypsy,
Hannah, Honey Terrence Myles, Hoté, Jake,
Jill, Jigger, Michael, Mimi, Minnie, Nellie, Nicky,
Perle, Precious, Rags, Rover, Saga, Sputnik,
Stevie, Tadhg, Tiger, Tove, Winky, and Zarko.

CONTENTS

9 *Dave*
Sara Baume

19 *Cats and Girls*
Niamh Campbell

31 *Light Thickens*
Darragh McCausland

43 *Millie*
Eileen Myles

49 *Fox and Sparrow*
Stephen Sexton

61 *Opening a Gate*
John Berger

69 *The Courage of Turtles*
Edward Hoagland

81 *What's New, Pussycat?*
June Caldwell

93 *Precious*
Jessica Traynor

101 *River before Me*
Tim MacGabhann

113 *Then, Horses*
Sabrina Mandanici

127 *The Roaming Edges*
Suzanne Walsh

133 *Pure Animal Instinct*
Vona Groarke

143 *Say God, Say Bird*
Latifa Akay

153 *Three Dogs and One Cat*
Honor Moore

163 *Michael*
Erica Van Horn

173 Notes on Contributors

Dave

Sara Baume

We are sitting in the yard between the back of the house and the front of the shed, my mother and me. There are islands of pot-plants in the weathered concrete and a flowerbed running along the wall behind us. It is summer and everything is in bloom – the geraniums, the lavender, the columbine, several spears of verbena, and a pillow of lithodora heavenly blue. There are windmills and other colourful plastic things that spin in the breeze on the end of long sticks protruding from the dirt. Above the flowerbed there is a trellis that supports a wisteria, a clematis, and an Albertine rose. My mother is particularly proud of that climbing rose. Apparently it is old-fashioned and was a gift from her mother. We are sitting on the bench with the yard table in front of us. On the table there is an embroidered linen cloth, a wooden board of cheeses, and a basket of chunky, seed-studded bread. My mother's garden is lovely in summer and we always sit out for lunch. We sit out in spring and autumn too, and sometimes even in winter. In spring and autumn and winter we sit out in our hats and scarves holding bowls of vegetable soup so thick you could turn them upside down and make soup castles. But today it is summer and everything is lovely when an enormous rook lands in the gutter of the shed and takes a moment to arrange himself comfortably and then peers down, enquiringly, at us and our detritus.

He is radiantly black, with a prominent beak pouch – this is how I know he is male. Males have pouches for carrying scavenged food around to impress the ladies, and then later – after they have been successful with the ladies – for collecting food to bring back to their young. All rooks have a patch under their chin that is bald except for a sparse, grey stubble – an adaptation that allows them to feast on carrion without getting flecks of blood in their black feathers. They are the ugliest of the crows but this one is majestic, with the sun glinting off his neatly folded wings. My mother, it appears, is already acquainted with him. He gets special rations, she tells me, a scoop of peanuts on the flat of the roof whenever he comes looking, which has become, for obvious reasons, increasingly often. They seem to have a thing going on – my mother and the majestic rook. He has become quite tame. We should call him Dave, I say, after my father.

I am standing at my mother's kitchen window; it is six o'clock in the morning. My mother wakes very early – three at the earliest, more often four or five. If it is three or four in the morning she will eat half a banana and neatly fold the peel around the other half and place it on the top shelf of the fridge and go back to bed. If it is five she gets dressed and has muesli. Without fail my mother is always up by six. This morning I have beaten her downstairs by seconds. I draw back the curtain to find a congregation of rooks – they are lined up along the gutter of the shed and the stone rim of the flowerbed; they are squatting on the bench, the table. They are waiting. They squint through the glass as soon as my face appears. Mum! I say, as she comes in after me in her dressing gown. There's a scene from a Hitchcock movie happening out there! Her hair, which was always short, grew long during the pandemic and she has left it that way. It is completely grey now and curls at the ends where it hits her shoulders. They're waiting for their breakfast, she says, and heads out the back door, barefoot, with her plastic scattering cup and a huge sack of peanuts. The rooks fling themselves down into the yard; they shout and flap. And there is Dave at the heart of it, unmistakable, taking angry swipes at his rivals who are too many, too fast.

I stay with my mother – sometimes for a night or two, sometimes for a whole week. I've done so with regularity since my father died and I am aware that when I tell people this it sounds as if I am doing it to keep her company, or to lend a hand with the domestic upkeep, but in fact my mother is less in need of company and help than anyone I know. My mother is constantly occupied – with her casual work as an archaeologist and in the archives of a local museum; with her garden, her reading, her granddaughter. She is so self-contained and solitary nowadays that I find it baffling to remember that she spent forty years cohabiting with a man. In truth, I go and stay with my mother for selfish reasons – as a hiatus from my own life. Sometimes she invents little chores for me. I wheelbarrow logs around or weed the onion patch or pick beans, but mostly when I am at my mother's I just sit around talking to her, or I go off walking the dogs over the barley fields, along the channels flattened by the tyres of gargantuan tractors. In my own life I live close to the coast and I always miss the sea after a few days at my mother's, and then I find that the barley fields give me a similar kind of fix – something about the purity of the horizon line, the undulation.

We are sitting on the garden bench beneath the macrocarpa, my mother and me, drinking mugs of hot water with mint leaves floating in them. The bench is several feet away from my mother's tiny wood. The wood is no more, really, than an untidy line of miscellaneous trees. The oldest is a Scots pine and the tallest is an ash and the youngest is a beech. Then there are a couple of sycamores. When

I was a child, the wood was unobtrusive but now the canopy is profuse; the branches grapple one another and overshadow the lawn. There has always been a rookery in the Scots pine, just a few big nests, my mother tells me, and Ruth who lives across the road had a similar-sized rookery in her garden. But then Ruth started to worry about her trees. They had become tall and unruly, and – unlike my mother's tall, unruly trees – Ruth's were overhanging the main road. My mother and Ruth live at a rural crossroads on the top of a hill. When I was a child it was bucolic, but in the two intervening decades it has become a commuter village for the city and the crossroads suffer a stammering flow of traffic, and because everybody sues everybody else over nothing nowadays, my mother says, Ruth became justifiably worried and summoned the tree surgeon who came with his ladders and straps and chainsaws and hacked all Ruth's trees back into gnarled stumps. And so her rooks had no choice but to cross the road and move in with the neighbouring rooks, and this explains how my mother's small, inconspicuous rookery came to be large, the dominant feature of the garden.

All over the grass, the hawthorn and the laurel, the fuchsia and japonica, the lupins and hydrangea, the redbrick path my father laid and the sandstone gravel he poured – there is crow shit. In wet weather it is slippery; in dry weather it is crusty. There is also a littering of bits of the things the rooks have carried home to eat – sugar-beet, apple cores, rock-hard husks of bread. For most of the year it is impossible to linger comfortably in this zone of the garden. In spring the shit is mingled with broken eggshells and lifeless fetuses. On more than one occasion I have wrestled a mangled blob of dead baby bird from the jaws of my scavenging terrier, alerted to its presence by the crunching of delicate bones.

When I am planning to stay at my mother's for more than a weekend, I bring my laptop and set up a makeshift office space for myself at a table in front of the window that faces the garden and the wood. While I am working there, I spend a lot of time gazing vacantly out the window. My mother has a bird feeder set up in the lower branches of a sycamore. She used to hang it on the washing line directly outside the kitchen window but then a rat embedded itself within the cotoneaster and loitered around the yard all day and so she moved it to the wood where the rat now lives happily embedded in the laurel and my mother conducts her BirdWatch Ireland surveys from the house through a pair of binoculars. Every time I glance at the feeder there is a rook swinging off it and at least three more poised on the ground, pecking up the spoils. They seem to take it in turns. The other birds – sparrows, chaffinches, greenfinches, goldfinches, wood-pigeons, blackbirds – hop sadly around the edges of the black fracas.

When I am not staying with my mother, I phone her on Sundays. We text ahead and arrange a time and so when she answers she has usually made a cup of tea and is sitting on the garden bench, or in the greenhouse, if it's raining. Wherever she is in the garden I will always be able to hear the noise of the rooks in the background. Rooks are the noisiest of the crows; all day they yak and cackle – quarrelling, cajoling, exchanging important information. They are louder than the cows, louder than the wind and rain, louder than the passing cars. Sometimes they are so loud that I miss what she is saying. From where I am, back in my life, I listen and I picture them: my mother and her obstreperous crows.

It is six years since my father was alive and since my mother became chief custodian of the garden. When my father was alive the lawn was always mowed; the hedges neatly clipped. The flowers remained within the confines of their beds and the vegetables grew in tidy rows. Nowadays the lawn is as much moss as grass, the gravel is invisible beneath the weeds and big clumps of the vegetable patch have been given over to borage and phacelia. The chaos is carefully ordered though. She shows me the drawings that she makes every new planting year in her garden notebook – each is an esoteric map of circles and scribbles and wiggly lines. Just recently Dave has started to tap on the bathroom window in the mornings, she tells me. He always seems to know the moment she gets up, and which room the bathroom is, and that this is the one she goes to first. He swoops down and taps on the glass and when she draws back the curtain there he is, sitting on the windowsill, waiting for her. And when she goes out to the yard to have her morning coffee he swoops down and sits at the opposite end of the bench beside her, and so my mother breaks tiny pieces off her slice of toast or her flapjack or her biscuits, and offers them to Dave.

She finds the noise of the rookery pacifying. It pleases her, she tells me, to see the rooks start to return early in the year. They go away to their roost in winter but by February she will start to notice them arriving during the daylight hours, clearing out the old nests and making repairs or building replacements. You'll see one with a big, awkward stick in its beak, she says, poking around, trying to find a place for it to fit. If a stick isn't exactly right, then they just cast it out. And then she collects the rejected sticks in her wheelbarrow for kindling, usually on Fridays when she babysits my niece. On Fridays I picture them – the oldest and youngest members of my family – together on the concrete floor of the shed stacking up twigs; building a human-sized, mountain-shaped nest.

We are crouching beneath the beech tree in the dead of night, my mother and me. She heard somewhere, on the radio, she thinks, that rooks snore. Apparently this occurs when it is very dark and they are at the deepest stage of their sleep-cycle, but it is almost impossible to hear because they will hear you first and wake up and alert the still-sleeping ones, which is what has happened now. The sound the rooks make tonight is a disgruntled squabble as opposed to a snore. Our knees are clammy with dew and the eyes of a rat flash from the hedge and out on the crossroads we can hear the voices of people walking home from the pub, laughing at a muffled joke. In the morning Dave is sitting – enormous, majestic, unperturbed – in the gutter of my father's shed. Pink petals cascade from the trellis beneath him. The morning sun illuminates his iridescent feathers – azure and emerald and plum. Inside the shed, amongst all my father's abandoned things, my mother's car is parked. She used to leave it out on the side of the road but since my father died, she always puts it away in the shed so that no one will know whether or not she is at home. This means she can hide if somebody rings the doorbell and not have her presence betrayed by the car. When we were children I remember hiding from the neighbours. We would switch the lights off and lie flat on our bellies behind the sofa, below the windowsill – my mother and sister and me – waiting for the doorbell to stop ringing so that we could get on with our lives, and it perplexes me now – when I think about it – that my sister grew up to be a perfectly well-balanced, sociable adult whereas I grew up to hide from people, exactly like my mother.

I try to make an effort – when I am helping her – to memorise the details of whatever it is we are doing – pricking out seedlings, building a bamboo wigwam, sorting the winter sticks – the why and the how of it. Someday I will be responsible for this garden – the vegetable patch and the greenhouse, the little orchard and the mossy lawn, the redbrick path and the sandstone gravel, the tousled hedges and overflowing flowerbeds, the wisteria and her beloved Albertine rose, the shed and its rusted, seized-up contents, the unruly trees and the tremendous rookery. Someday all this will be mine, even though I do not want it.

Cats and Girls

Niamh Campbell

In 2017, the Metropolitan Museum of Art refused to concede to an online petition calling for the removal of *Thérèse Dreaming* (1938), a painting by Balthasar Klossowski – known as Balthus – which was included in an exhibition entitled *Cats and Girls – Paintings and Provocations*. The work shows an austerely suntanned twelve-year-old (Thérèse Blanchard, the artist's neighbour) in a pose that is relatively comfortable for a Balthus composition: she reclines and closes her eyes, raising her arms. A leg hitched on a table lets her skirt fall back to reveal the gleaming gusset of her underwear. The petition claimed that the painting sexualises a child; the Met countered that the work's ambivalence should not be reduced to that, or censored. A press release for the exhibition itself frames the several paintings of Thérèse on display as evidence of Balthus's 'kinship with children', a swerve towards the desexualised, which accidentally approximates some of the queasy affect surrounding the singer Michael Jackson. Beholding *Thérèse Dreaming*, I think it's pretty obvious that the painting is intended to be erotic. Of course it is. But who's pleasure are we seeing: who sees?

I confess to finding these kinds of insouciantly open-ended questions irritating these days. But when I was a graduate student I talked like that a lot. *Deconstruction* was frotting noncommittally at the edges of *discourse*, like a child tearing wallpaper down behind its crib, as I also enjoyed doing once; ideas of otherness, openness, speculation, and a kind of unforeseen-because-tentacular-and-notional solidarity inching into the ether appealed to me. This is because I understood what they meant, and these days I no longer understand. Or else I poorly understood them the first time round, I was just breezy and brimming with youthful self-esteem. These days I feel less complex and even, sometimes, like an intellectual sellout, or dropout. I feel this when I do things like toss a plastic bottle into a bin. I think of a video I once watched of the academic Avital Ronell modelling moral thoughtlessness by singing 'I'm proud to be an Okie from Muskogee' at some kind of panel discussion.

At the same time, I also retain a dim sense of potential in asking *who sees*, a reflex drawn from the gregarious, wrist-tossing, type-as-fast-as-you-think frilly anguish of the doctoral dissertation, not because this reflex seems to blame the viewer – you – for eroticising Thérèse, but rather because there is another set of eyes in the painting itself. On the floor at her feet, lapping slenderly at a saucer of milk, is Balthus's second favourite subject: a cat.

Mitsou, Balthus's first cat, arrived into the painter's life when he was ten, only to run away again within a few months. In mourning, young Balthus made a series of ink drawings, *Mitsou*, which were reproduced and disseminated by Rainer Maria Rilke, a lover of Balthus's mother. The boy Balthus grew up in an affluent and enviably cultured émigré enclave of Paris, and later styled himself as the Count de Rola. Bono sang at his funeral. One morning, recently, and out of basically nowhere, my boyfriend messaged to ask what I think of Balthus and to express amazement at the fact that Bono sang at his funeral. Which really is, when you think about it, weird. Balthus reads in retrospect like the kind of person you'd like to see punished, but who went entirely unpunished in his lifetime. *Mitsou*, all the same, inspires tenderness.

In these drawings, young Balthus, in short pants and with a pudding-bowl haircut, cuddles and scolds Mitsou; they read in bed together, gather with family around the Christmas tree. When the cat takes off abruptly, never to return, the boy is left bereft, crying alone. Where the cat appears in a panel head-on, its facial expression has that insular, indifferent quality of all cats' expressions, implying repose to the point of somnambulance, and the work is so good, so suggestive of incredible talent, that this muted cuteness is especially pleasurable. The cats in Balthus's adult paintings, however, tend – for me – to be consciously non-realist, simplified, and sometimes demonic, their grimacing faces more like tribal masks. There are so many of them across his work, cats and girls cavorting, girls like cats in four-legged positions, a cat leering out of the darkness at a naked girl. Cats just *there*, on carpets and under tables, like actual cats. The press release presented by the Met for *Cats and Girls* suggests these felines represent Balthus himself (in his twenties, he painted himself as 'the king of cats') but this is, to me, too literal. What is a cat in life, in relation to our life? Another question, pressed to you like it is urgent: your problem now.

I don't own a cat, but I know one well; we hang out frequently. A plump tom who is now also obliged to hang out with a baby, the most recent addition to my sister's household, and to bear with grace the pokes and pulls and attempts to stand on him undertaken by this baby. But instead of staying out of her way – the way of my niece, Lani – the cat comes and sits next to her purposefully, looking distantly over her shoulder as if beyond agenda, like Mitsou. He walks over her, bumps his head into her, and finds reasons to lie close to her as both of them move through the rooms of the house like sunlight over the course of a day. He waits close enough for a poke or a shove, long enough to provoke a poke or a shove, and then, with this excuse, he swipes a quick and declawed paw across her face. Many aspects of this are worth remarking on – the proximal trolling, the conscientious folding of claws. He wishes, we assume, to be permitted to hit her, but only benignly. The claw retraction may be kindness or a strategy to ensure no injury sustained is ever bad enough to deprive him of access to the game. It's true that when I watch this – with little sympathy for baby or cat, both of whom know what they're doing – I feel the animal is letting himself down; in betraying this petty need for tiny acts of vengeance, he seems human. Understanding his own power in the household to be irreversibly diminished by Lani's arrival, he can only console himself with malice: understanding the cat to be the one individual in the house ranked lower than her in terms of size and strength and liberty, Lani can only console herself by tormenting him. Their tension is, finally, a kind of love.

On the internet, several photographs document a period of time in the 1950s Balthus spent living, in advanced middle-age, with his seventeen-year-old step-niece Frédérique Tison. One striking image shows Frédérique, her arm arched upwards in a pose similar to the one held by Thérèse, petting a scruffy black cat. I think these images are the reason why someone once told me erroneously that Thérèse of *Thérèse Dreaming* was Balthus's niece, confusing the muses but also invoking a kind of perv continuum. Muses as interchangeable, concertinaing paper dolls by American Apparel. To think this is to feel insulated by cynicism, though it is not necessarily true. I remember some *Guardian* journalist or other, addressing the Woody Allen phenomenon, perhaps – *dirty old men should keep their hands off fresh young knickers*. 'Knickers' being a word that possesses, of course, that BBC Home Counties note of menacing self-infantilism; it is up there with 'spits and spots of rain'. Turning a schoolmarmish tut-tut on the old men feels as dissatisfying as claiming Balthus felt kinship with children, or insisting that actually She's Really Mature for Her Age: we are all of us adults, not Okies.

Redeeming *Thérèse Dreaming*, Lauren Elkin defends the eroticism of the image in a way I find myself essentially agreeing with, against the petition that circulated in response to the Met. 'I've always loved this painting,' she explains (so do I); 'It makes me feel sexy, but in an adolescent way; it restores to me, retroactively, the diffuse desire I felt at 12 or 13.' This desire was nebulous, not directed, and really – as I understand Elkin, as I remember being a child myself – a sensation on the cusp of innocence and knowing, of auto-eroticisation as a hermetic, basically antisocial *event of the embodied self*. It stood in opposition to the other kind of auto-eroticisation I was learning as a child, which involved realising I could hold my body in certain classical or modern poses and elevate it instantly to the level of desirable. I knew this as young as twelve, maybe. I have a photograph of my little friends and me, aged ten or eleven, posing with our medals on the front lawn of a suburban house, and one of these little girls is inclining her head and pouting suggestively. She used to pull this pose in photos so often we started to mimic her. I look back now and wonder what she was exposed to, or whom, in our fretfully protected Okie universe of lawns and medals to have been so ahead of her time.

Thérèse's luminous eroticism might, then, be reclaimed as an inadvertent celebration of that moment in a little girl's development right before your sexuality is identified, vilified, recoded as criminal, and taken away from you – right before you turn into jailbait without noticing, right before the hostile snapping of adults to close your legs and put on something longer, that horrible vehemence that is bewildering for a time, then finally not. I remember this state also, this stage of life in which I read library books on beach towels and rockeries, wrote stories about Victorian ghosts, and learned to masturbate with intense discretion so my mother wouldn't hiss at me to stop doing that. I had no interest in boys, still less in those shadowy creatures *men*, apart from my daddy, onto whose knee I chastely climbed to watch TV. I found the drama of *Malory Towers* highly arousing. I had posters of ponies and prayer cards on my bedroom walls. And I hate the thought of any precious part of this being converted to titillation for anyone else; I do not like to think of the thrumming, puppyish bodies of prepubescent girls becoming fetish fodder; I do not like to think of myself endorsing this, like a Cool Girl. And yet – *and yet*, old ambiguity is the tool I have to reach for here, dissertation again – and yet it seems equally annihilating to object; I am struggling to articulate this because there is no register, in fact, in which I can express girlhood without reverting to *the girl* as trope and product and object and thing. And so it may be easier, more diffusely accurate, to try and talk about the girl as cat.

Frédérique petting the cat, with her long hair drifting down, looks like a schoolgirl witch. These cats in Balthus paintings don't, for me, stand in for Balthus, nor for *pussy* even – far too crude – but for a form of witness, a seeing, which is detached from the human but still addresses the human. It watches across a gulf of incomprehension, its gaze does not encode curiosity or fear or love or anything but a sublime, planar blank. In my student years I read Derrida's *The Animal That Therefore I Am* (2006), in which the philosopher reaches something like this conclusion about looking at animals – being looked at by animals – by spiralling out from the scenario of standing, naked, in front of his cat. The cat looks back, but what is the meaning or measure of this look? The impossibility of knowing is a kind of humility. One day last year, I put my niece in a static walker, watching *Elmo's World* on a laptop, so that I could take a shower. As I was towelling off I heard a plastic clatter from the lower room. I feared the worst. I rushed down naked. Lani had tossed her plate of toast at a wall; as I halted before her, her face opened with incredibly tactless drama in shock – her eyes widened, her jaw dropped open, and she froze. The cat, unperturbed, watched me glassily from his loaflike pose among the sofa cushions. It occurred to me that clearly Lani had never seen a fully naked individual and that this might become her first memory, such that I would one day need to explain why I loomed nude over her when she was less than two. I felt embarrassed. The cat, as per Derrida, is always naked, or some form of naked, however that is experienced by it: a theta-level sense of self as network integrated into another networks – the sound of birds outside, the soundless swing of the gate he always detects first, fresh air and decay, forms of human surplus like sebum or cell debris glowing subtly on the furniture, like bloodspray under ultraviolet light. This animal is entirely concerned with itself. Our passionate attachment to him lies somewhere between condescension and reverence.

Some people love animals because these animals, as their owners see it, serve them, receiving projections and returning pure dependence with a loyalty that is total (dogs) or spiced enough by aloofness to appear elite (cats, horses). Culture – the kind of culture that produces Balthus paintings, for example – also loves girls because they appear to serve or service, to be passive, to be pettishly aloof but powerless. Power can only operate within the system of objectification into which the girl arrives as soon as she becomes aware of herself as *girl*. This power is seduction, manipulation, pretend innocence; it brings actual liberties and rewards. It brings knowledge. At its best, Frédérique or Frances from *Normal People* or Nicole Diver or Marie Calloway or Lolita or all those girls in those Fellini films, those Woody Allen films, that Philip Larkin fantasy of boarding school – girls, to quote Joanna Walsh's *Girl Online* (2022), 'threatened by their own exorbitance' – have a lark and move on and don't get caught in the dark, transfixing gaze of self-objectification as suicide. They find their way back, as I did, to the bleak core of their own unrepresented, never-to-be-represented, selves: the dream of Thérèse, humming inwardly. I can remember being that age but I cannot tell you about it without entering a language of soft pornography, because soft pornography has come for my – for your – tender youth. As a form of knowledge, it can only remain preverbal and approximate something of the swing between boredom and arousal exhibited by the purring cat; a creature we do not understand who nonetheless has been placed into a framework of interpretation. Those things people hate in cats are the same things people hate in girls, incidentally. They are killers, disloyal and absurdly elegant, they are not really the fluffy fools we think they are. People tell you this as if you've been taken in by a cat, you must be released from the thrall of this cat, you could not possibly admire the impersonal pleasure-seeking, the flopping onto couches and stretching limbs idly, common to girls and cats. But even that is *objectification*, actually, and I want to end on something closer to the truth.

In my phone I have a photograph of Lani, taken moments after what we called *the naked tantrum*, an hour-long dilation of rage and refusal that began out of nowhere after we had stripped her for a bath. She howled and raced around the house, peed on the carpet, and was finally felled by Calpol, lapsing into a sulking languor and lying on her back. The photograph captures a scowl, hands resting on distended infant belly like a frog, and the large pupils of her eyes give back a felinely unyielding but accusatory stare. If the cats of life as well as those of Balthus bear witness, representing a gaze within the gazing space of the painting, we are subject to this neutral witness without recourse to any final interpretation: Am I judged? Am I understood? Am I instrumentalised? Watching a small child be inducted into socialisation provides a glimpse into the animal reluctance we must all have felt, at some point, in the face of language, the symbolic order, etiquette, reification, rules. Lani doesn't throw naked tantrums anymore because she has this *under control*, but precisely what is lost to control is the same as the simmering subjectivity we can assume is present in Thérèse, behind those eyelids, an inexpressible and wholly inaccessible, divinely private libidinal state, before anybody turns her into anything.

We might consider it the origin of dignity, really, as well as savagery. Frédérique looks so happy with her cat; Mitsou, the original mystery, takes off on a whim into the Parisian night. Thérèse, apparently, dies at twenty-five, leaving no commentary.

Light Thickens

Darragh McCausland

As my ageing father's Skoda trundled over a cattle grid, every bump tearing through me, my mother said, 'Look at those happy men.'

There was a determined and cheery note of hope in her voice. It gave me the impetus to lift my head, which had lolled between my knees for the two-hour trip south to Ireland's addiction treatment centre for men. A last-chance port of call for those about to permanently fall off the map. Battling the instinct to vomit and the fluid shapes and angry sounds that passed between my mind's eye and ears – the results of alcoholic hallucinosis – I heaved my giant stone head to its natural position. I rotated its mummified face sideways to stare at the creatures in question. The happy men. I felt immediate horror.

There were three of them. One leaned on his rake and waved. The other two were paused in the act of picking up foliage or sticks. All three seemed to grin towards me in the same semi-lobotomised fashion as the fireman who waves in slow motion from the eerie, sun-bedazzled 1950s fire truck at the start of David Lynch's *Blue Velvet*. I licked my puckered, papery lips and felt my spit as off-putting as envelope glue.

'Ffffuck,' I said and let the stone head drop with a suddenness that was the exact opposite of the force with which it was lifted.

I oxygenated myself with the whooping breaths that were my only means of staving off vomit. The car slowed. My father rolled down his window to talk to another (presumably also happy) man in a security hut. A breath of early summer air touched the sweat weeping from my neck. Somewhere beyond the car, beyond the noises in my own head, I heard the chatter of many rooks. I had arrived at what was to be my home for the subsequent thirteen months.

Eight hours later, I was clad in detox-requisite paisley pyjamas and slowed on a forty-milligram dose of Librium, a drug as familiar and comforting to me as it is to any alcoholic who has had to 'white knuckle' the mania of withdrawal without it. I had been placed in a small room in the detox unit with a puppyish, lamb-haired cocaine addict who was fourteen years my junior, and who we will call D. We exchanged a sort of punch-drunk, befuddled chat about how we both ended up in that little monkish room containing two beds, two lockers, and a strangely moody picture of the Virgin Mary that was far closer to Pasolini's version of her than any representation I had previously seen on an Irish wall. She seemed to look at us as if we were a puzzle, rather than creatures upon which to pour her infinite mercy.

Being a certain stripe of cocaine addict, D. was a doer. At that early point he was already engaged in the primary activity that was to occupy him throughout his eight-day stay in the detox unit: counting the individual pieces of the many raggedy one-to-two-thousand-piece jigsaws and sealing them into sandwich bags. I think he did it for two reasons. The most obvious was the residual obsessiveness of a stimulant addict. The other, the reason I now think motivated him most, was that he simply had a great big heart. I guess he did not want to experience empathically the disappointment of the other men around us trying to finish jigsaws missing their final few pieces. At a certain point, after a golden diamond of light on the bedroom wall had faded to cream, he said to me, 'You're sweating bullets, pal. Get outside.'

There were three large wooden benches lining the wall of the detox unit, all painted a strong blue that I would much later discover (when I began painting things there myself) was a Dulux-brand colour called Killala Bay. When I sat out on one of those benches, I was in a ferociously unfamiliar place. My mind at that point had only the most tenuous grasp of how large the entire treatment centre and all its zones of activity were – its gardens, a farm, a carpentry centre, turf sheds, little villages of houses for long-term residents, and a rookery. Now, when I look back on that moment that was reduced to an indoor bedroom, a blue wooden bench, and the sight of the rookery ahead of me, I wonder at the absolute strangeness of the way we retroactively project space into time. How a complicated and vast environment that took weeks, even months, for me to grasp and map, now automatically seems it was always so, but only so long as I do not revisit that initial moment that undoes it all.

The Killala Bay paint was flecked with the acidic droppings of rooks. In what was maybe my first act of self-care in a year, I was mindful about where I sat. I let the heavy book I optimistically cradled (let's call it *The Brothers Karamazov*) drop to my side. I gave myself over to the prospect in front of me. Made up of a linear copse of pine trees, the rookery also served as a rook-roosting colony in the off-season. But not only that, I later found out that it was one of the largest and most established in the centre of the country.

My nerves were simultaneously aflame from withdrawal and packed in a spiritual shoebox of faintly blue-coloured dense benzodiazepine fluff. When the rooks announced themselves properly, it was into that contradictory space. I happened to sit outside at the precise time of day when they stream back across the fields to their home. First in single lines, then in chains of dozens, hundreds, and, at the very end, blackening the sky in wheeling thousands. Streaming, with harsh cries, they made wing home from the eight points of the compass. An unease overtook me as I watched them pour and wheel across the sky. It had less to do with the birds themselves than it had to do with the place they had made their home. I was not able to put my finger on it, but I figured that there was something unutterably wrong about the scene. Sweat filled my creased eyes and blurred my vision, exacerbating my alarm. A corresponding drop of mucus formed in my nose, and I wiped it on the sleeve of a sweater I bought in Topshop eighteen years prior, which was precisely around the time certain friends began to say variations of the statement, 'Drink does not suit you, Darragh.'

I did something very weird. I opened my mouth. Words burbled out.

I said, 'Light thickens; and the crow makes wing to the rooky wood.'

*

I have a non-academic apprenticeship with the English language literary tradition. At best, it's impressionistic and groping. At worst, it's sloppy, lax, and conveniently dishonest. Either way, I have a loose and subjective connection to the reality of the canon. With this caveat in mind, coupled with my perverse refusal to look up something very simple in case it wrecks a thought to which I've clung, I'll state here that 'I think' the American poetry critic Helen Vendler once made a very profound statement about the poems of Wallace Stevens. 'I think' I heard this in a lecture I watched on YouTube. Here she comes, in my mind now, stating what she may or may not have once said: 'Wallace Stevens's poems are like Christmas trees that are initially bare but decorate themselves with meaning as years go on.'

If she did indeed say this, I think she meant that a beautiful line of poetry might lodge itself in a receiving mind at the initial point of contact without transferring much meaning beyond aural beauty, hook, potentiality. After that, the mind holds it dormant, like a seed, waiting for a future event to water it or pour the catalyst on it that activates its meaning. It's a magical idea.

Studying Shakespeare in secondary school, I went to see a production of *Macbeth* in Greater Victoria, Canada. At that point in life, I was in a state of permanent dissociation, and the poetic meaning of anything beyond the melodies of songs was wrapped in shadow or held at bay behind the coruscating fly's eye of mental mirrors, which reflected nothing back to me that was not the narcissistic construction of a certain type of lonely teenager's self. Yet a piece of poetry from the play penetrated me and has stuck with me ever since. It is Macbeth's murderous incantation where he invokes nature to sympathise with the deed he instructed his accomplices to carry out:

> Come, seeling night,
> Scarf up the tender eye of pitiful day,
> And with thy bloody and invisible hand,
> Cancel and tear to pieces that great bond
> Which keeps me pale! – Light thickens;
> and the crow
> Makes wing to the rooky wood;
> Good things of day begin to droop and drowse,
> Whiles night's black agents to their preys
> do rouse.
>
> (III, ii)

Although I gave bare thought to the endless depth of their meaning, these words fairly sent a shiver up me, and roosted in my young consciousness in a way that was to prove permanent. In the years from then to now, I have repeated them to myself variously, sometimes sober, more often drunk. Drunkenness more than sobriety seemed to bring up darker poetry, baleful lines that would announce themselves on my lips as impenetrable mantras clouded with jeopardy. And I would always say them out loud. Every single time. Like Samuel L. Jackson's character in *Pulp Fiction*, Jules, reciting Ezekiel 25:17 before assassinating a person, because 'it was some cold-blooded shit to say to a motherfucker before I popped a cap in his ass'. Except, in my case, I was the ass I popped my own cap into.

Macbeth's words returned a second time sixteen years after I first heard them, when I wrote them into a short story called 'Grotto'. In that story a bewildered boy sits in a psychiatrist's waiting room that has one piece of art on the wall: a photograph of a crow examining its own foot. He is interrogated by the psychiatrist on account of leading a mass hallucinatory incident in the fourth class of primary school where a dozen children see, after weeks of staking out the trees at the back of the school, a green-faced demon perched in a high tree. To write the scene authentically, I tried to become (re-become?) the boy. I propped myself against the back of the bed on which I wrote and promised I would write the first strange phrase that wetted my actual, human tongue.

'Light thickens.'

Macbeth's speech suggests a wrong turning in the stuff of reality, a curdling of the visual environment. On the evening I sat on the shit-flecked bench outside the treatment centre, there was much wrong inside of me. Because I was in a state of hallucinosis, where two-dimensional things run like three-dimensional fast rats, I could not be entirely sure if there was something wrong with the rook colony in front of me, or whether it was a mental trick.

Two days later, I worked it out. The aspect of the scene that was wrong belonged to reality. It belonged to the shapes of the tops of the trees. They did not rise vertical as nature intended. They rather drooped sullenly as if they were made of wax brought close to a source of heat. In that regard, they were like molten birthday candles. How was that so? What force had made those trees that drooping shape? Distracted by the thought, a range of other blackened thoughts followed fast, the varied and harrowed fears and anxieties of deep addiction. Like birds, they settled in me. As they did, I felt for the first time a shaping weight, a pressurising, ossifying weight. The outer environment chimed with my inner state and an answer of sorts did too.

Entering stage left, a rook settled on the highest drooping pine top. Another followed. And another. When the three perched together, I perceived the supple wood give way under their combined weight. Of course. That was how the trees had grown wrong. They had been altered under the weight of the rooks that settled on them like my own dark thoughts, my repeating, strange, barely understood thoughts that had sat in my mind since the first time I picked up a drink. My brain was the trees, permanently altered by what sat on it over time. A slave to visual metaphor, I looked at a couple of poppies that trembled in front of me in the darkening aspect of wind and light. Me again, I thought. Meanwhile, the other men in the detox were indoors, playing Monopoly together, laughing loudly. I had refused all their invitations to participate.

*

Every one of us in treatment was given a compulsory job in the centre. It had a public-facing restaurant.

My initial job was to work as a line chef in the back kitchen. It was a position that appealed to me in all sorts of ways, not least in my absolute willingness and masochistic desire to submit to the orders of others. I see now that this is a part of my addictive personality. The allergy towards responsibility.

By the time I began working there, I had stopped sweating according to my nervous system, though I still sweated according to the heat, for the summer turned out to be bright and strong. With that physical improvement, I had learned in the meantime to write words on paper without it looking like a crow had stepped in ink and hopped across it. The psoriasis that had covered my entire torso had cleared up. The teak colour in the hollows of my eyes had turned to something approaching flesh. Yet, more remarkable than those physical improvements, came the headway I made in my interactions with fellow humans. I was no longer a paranoid atom. There were even one or two men I sought out in the evening when the time to relax softened around the entire house and its one hundred residents. I still found small talk hard, mind you. I had to hinge it on a subject that obsessed me.

'Those crows are some operation. There is a lot going on up there with them,' I'd say.

Different guys in the house would have their own ways of responding to the gambit. I started to love how their answers always revealed something true about their own peculiar natures. A craggy man in his sixties who had never revealed his homosexuality to his nonagenarian parents wept at the end of a conversation that began with the specific roles of crows and ended with the battered shape of his heart.

Another man, florid in his ideas, told me that 'the nun who runs this place controls the crows. They protect her', and afterwards pressed soft paper that looked like a shopping list, his handwritten cure for the Coronavirus, into my hand.

'Are they like the winged monkeys?' I asked.

'Exactly that.'

I laughed, because the nun in question is the most deeply serene and gentle soul I have ever met. This man was gentle too, and the fact that there was no incompatibility between his words and how she received him personally, taught me something profound about what it means about being a truly tolerant human, as in, tolerant of all.

I stayed at the treatment centre for the best part of a year, volunteering as a supervisor in the craft shop. These days in Dublin, as the temperature drops, and leaves are torn from trees in thickening light, I occasionally look at a crow making wing (to where?) and think of the men I got to know in that strange and special place. Not without pain, as a few are now no longer with us. Recovery is full of fellowship, colonies of people chattering, looking out for each other, co-operating, like crows in trees.

Millie

Eileen Myles

I'm glad I have a bad dog. Earlier tonight I saw on Twitter that the City of New York killed Millie. She was a pale honey-coloured dog, two years old, slight looking but probably some kind of pit bull with large floppy ears. I put money down on all kinds of dogs. I only realised maybe a year and a half ago that besides fostering or adopting a dog or doing nothing at all you can pledge. That means you put ten or twenty dollars on a dog in a shelter. The dog's dowry grows and by the time a rescue takes the dog they have several thousand dollars. Sometimes a dog will have a huge dowry and the shelter will kill them anyhow. Because someone has to actually rescue the dog. Meaning take them out of the shelter. And you look at some dogs like I looked at Millie and thought, this is a cute dog. Someone will save Millie. It's Ace I'm worried about, or Bailey. Or Bone. Or Gumbo. But Gumbo got saved and Millie is dead. It's possible there was some little item in Millie's story. Like on Twitter it even said she's a sweet dog, she just needs structure. So apparently she had a family or an owner and they didn't train her and she did something bad. And now she's dead. And they don't kill dogs in nice ways. It's not like when they tra-la 'put your dog to sleep'. I think they inject some horrible burning City of New York chemical into the dog, who may or may not know what's about to happen. I'm walking down the street with my dog, Honey, and I'm thinking about Millie. I read about Millie early in the night and it flavoured my entire night. I felt sad. This is a sweet dog. I can practically feel Millie in my lap. I can go and find videos of Millie now showing how sweet and cute she was. Now she's like a piece of trash in a bag somewhere being brought to the dump. I love to tell the entire story of my dog, Honey, but I'll cut to the chase. She was in and out of the shelter twice, which means for whatever reason she was abandoned, left on the street, while young and/or brought back to the shelter at least once. Those dogs have no time at all. A returnee has like a week. Maybe. Think of a culture that lets dogs just wander the streets. They don't pick the dog up. The dog exists on scraps. Maybe someone adopts the dog in some way for a while but nothing permanent. In the state of late capitalism however we can't let dogs just wander around. It must be owned or captured and put in a cell and if the dog is already proven to be flawed – unowned, or untrained, or quick to defend itself, or doesn't want to leave its cage – it must be killed. It doesn't even have the right to breathe the same air as us. It must die. Only good dogs and dead dogs. Honey liked to put your hand in her mouth. Honey liked to leap up in your face, friendly like, but it could leave a couple of prong marks from her teeth in her big smiling happy face. That happened to me at least once and I bet these kinds of things put her back out on the street and back into the shelter. And I thought when I saw those little red prong marks under my nose, I have a dog that bit my face. I felt scared and I saw her huge energy and I thought this is unmanageable. I had thought a little while back before I got Honey that my next dog would be a really easy one. A dog that doesn't challenge other dogs in the street. That is compliant and friendly. An easygoing guy that everyone loves. My dog acts like she will kill little tiny cute dogs, she will attack an older frail dog. If she has the freedom to run out the gate of her yard she will attack perfectly nice dogs being walked on a leash. I was told by my ex I love calling her that I was told that people in that town where we met don't like my dog. I felt hurt. That people were talking about my dog in an unfriendly way. Because she is my dog after all. I decided to keep her. I don't know how. I don't know when. In the midst of the badness I saw sweet and good. I saw beautiful. I saw needy and abandoned. I saw left alone all day and yelling like hell. Bark bark bark. In there somehow I thought I will not return her. I knew that if she went back to the shelter they would kill her. I walk the streets of this city and all cities and see dogs being walked by their owners and in most cases I don't think they are any different from the dogs in cells getting ready to be killed by the city. The city is capitalism, nothing else. It cares about something and it cares about death. It cares about goodness. It keeps the good dogs and it kills the bad ones. It sometimes makes a mistake but that is its intention. To separate the wheat from the chaff and to kill the chaff and save the wheat. My dog is beautiful. My dog has mad-crazy eyes. Deep beautiful wild-animal eyes and calm awareness of her own beauty accepting your admiring eyes. She can be a beautiful piece of animal furniture looking good on the mat or on the couch in the house. She is pleased by her surroundings and she accepts them. She wants a home and she has it. She lives with me. When I think of a dog like Millie that was killed, pointlessly and unnecessarily, and I think of all the dogs that have homes and all the dogs that don't, I think the fact that I have managed to coddle and feed and walk and spoil and love and sleep with one of the other kinds of dogs in the world, a smart dog, a shrewd dog, a quirky dog, a pleasure-seeking dog, a specific dog but not a good dog. A dog that randomly makes a lot of noise. When she leaves one city in a vehicle she barks. When she is coming down the main drag of the new city where she is arriving she barks again. When she sees dogs in the street before she's even gone home she barks like she wants to kill the other dog. People look frightened and wonder, looking stricken and even outraged, what's wrong with that dog. And I feel like saying fuck you. She is just barking. That's what she's doing. Sometimes other dog owners look and smile and laugh. My dog is bad too. Or that my dog is just being a dog. And it's true. I don't know if I ever had a real opportunity to make my dog into something else, someone else. I tried a little. I got involved in a whole cult of people who lived in squats and did sex work and they really knew how to train pits. The guy sat on the couch right next to her and lifted his hand and she stopped barking instantly. How did he do that. I never learned. He advocated collars that would give her a little shock. That just seemed wrong. My friend Alice asks why I don't get her a muzzle. But Alice's dog is bad. Why doesn't she get her own dog a muzzle. That seems medieval. Like marking my dog like those women who would walk around medieval Spain with golden masks on their heads and inside they were starving. I would not do that to my dog. I must stay awake and keep her alive. We've done seven years so far. I thought when I got her I will be so old when she dies. I am half that old already. She is older too. We are getting old together, the bad dog and I. She is lying on her bed eyes closed, legs leaning on one another, paw extended, she opens her eye for a moment. She can feel me looking at her. She returns to her state. I will go towards my bed and she will rise. She will place her paws on the bed and I will lift her rear and she will hop on and take my spot and I will go uh huh. And she will shrink and reposition herself lower on the bed. I have mastered that.

I think then of the larger unkindness of the same killing city that wants to take the large park away from us and we are the animals that use it. And I will probably write more about the animal nature of my kind but this is all for now.

Fox and Sparrow

Stephen Sexton

Because it was almost Christmas, Ciaran, my dad's friend of many years, left two gifts on the kitchen table and settled himself into the black sofa across the room, at a right-angle from my dad's mint green two-seater facing the ever-running television. It was December 2019.

For thirty years or more, Ciaran has been delivering to my father the tokens his increasingly elderly mother – time happens – sends from where he lives with her in Drumaness. He was a welder for forty years; my father retired a decade or so ahead of him after a similar life as a crane driver. They sympathise with each other about their COPD. Treacherous working environments: no masks; no knowledge – in those days, they said – of what they were breathing in. They both smoke enthusiastically.

Most years, Ciaran arrived around noon on Christmas Day, not yet *rightly*, as he'd say, but quickened by a few drinks he'd taken at his brother's house, while the nephews inspected their hauls. Ciaran didn't drive in those years, so his brother took him an hour on the quiet roads to our lemony pebble-dashed, semi-detached house on the main road just outside Ballygowan, nine miles from Belfast. Countryside: farms, cattle, crops; in winter, the bare trees of winter. When my mother died in 2012, he stopped coming for Christmas – my dad and brother and I constituting only a house of men – but visited instead in the third week of December and stayed an hour.

Even before the pandemic made isolation general, my dad spent long periods alone at home, unsteady on his feet, hard of breath. I'd just moved to Derry. My brother was teaching English in China or rambling around Thailand: a kind of grieving, I suspect, though we never discussed it. More than once, my dad mentioned loneliness. It was only him in the house now, called in on by a loose schedule of carers. Loneliness was a severe, smarting disclosure I did my best to soothe before starting the car, twisting on the big lights and pulling onto the road.

That Christmas, in 2019, their conversation was directed by the television, and they talked to each other by talking at it; a shy, refracted speech. Since darts was on, they talked about darts: which pubs had what kinds of oches; where the dart board was situated (and how foolishly); who threw well and who had no hope. They played on a team together. By extension, they reminisced about which Saturday night dances they used to go to in Ballynahinch or Downpatrick and – this being the 1980s – which forenames they'd inhabit for the night depending on the part of the country they were in. *Ciaran* might be conspicuous at dancehalls in certain parts of Antrim (those parts of Antrim had dancehalls), so he'd go by another name for the night.

John Montague called the dancehall scene 'an industry built / on loneliness'. The ordinary aches of unhappily single people. For my parents and their friends and for other pseudonymous goers-out, there must have been a peculiar and intense loneliness to these Saturday nights. By taking a new name, to mitigate the risk of suspicion and sectarian dispute, they were estranged even from themselves, if only for a night; separate from their names and their political and religious associations. And they were just people in a hot room with a showband, trying to kiss someone in the car park.

Just the way it was: changing names, checkpoints, fear; a way of living almost inconceivable to me. They were all around County Down and elsewhere. They were in Loughinisland from time to time and knew some of the people massacred there in 1994 by the UVF, allegedly in collusion with the RUC. They gave those people lifts home down the dark country roads with which the past is riddled. Then my dad told a story.

*

When he was about sixteen, in the mid-1960s, and living at the family home near Rathfriland, my dad made a bit of money hunting foxes. They had no electricity or running water, and a big family. They trapped rabbits for the pot and planted and picked potatoes for farmers in the surrounding fields, who'd give them a sack to take home sometimes. All upright Presbyterians were the farmers. And, my dad would add, most of them were nothing but kind to his family. There was, for the most part, mutual respect when it came to the differences of rite and ritual. If someone died and there was a wake, the farmers would come to the house, but they wouldn't enter it; a propriety more familiar to me from Seamus Heaney's poems than my father's life, which, oddly, therefore, sounded more like cliché than fact. Some evenings on the way home, my dad would get the news from one of the farmer's radios. This is how he learned Kennedy had been assassinated. 'Tell your father they got that boy of yours today in Dallas,' my dad remembers a neighbour telling him in November 1963.

There'd been a bounty on foxes in place in Northern Ireland since 1943, instigated as agricultural pest control. It ran until 1977. Farmers sought help defending their poultry and lambs from opportunistic foxes. In morbid, noirish prose, the naturalist J. S. Fairley, writing in 1969, pathologises the animal as 'wantonly destructive', biting the heads off fowl; killing more than they eat. In a wonderfully vampiric close-up of the savaged hens, he describes how the 'marks of all four canine teeth are frequently visible thereon'.

Fairley also reports that foxes were thought to return to the same farm, in acts of serial mayhem. In the months of May, June, and July, he even proposes 'family parties' of foxes might carry out raids together. The fates of lambs are gruesome too: damaged skulls, missing jaws and heads, punctured throats. According to other studies and papers published in the 1960s, foxes have been known to open the chest and abdomen of lambs, as well as attacking the jaw and tongue. Not all foxes kill lambs, Fairley would have us know. Rather, certain 'rogue individuals' are responsible, with the suggestion that these foxes may have been weaned on lamb, having therefore developed a taste for it.

Like lambs, fox cubs are born in the spring, though for the purposes of the bounty, they were considered adults at the beginning of July. When the young are grown enough for solid food, a few months later, the father fox goes hunting. This was when official hunters were engaged, and 'skilled amateurs' too, as the naturalist describes them.

My dad was out one day with fellow skilled amateurs: his brother Brian and Pat McGinn, a man without a tooth in his head, and their terrier, Whiskey, sniffing out a particular fox, which, after some combing and topography, they tracked down to an earth at the edge of the farmland. The dog frightened the fox enough to make it bolt; the men grabbed it. Usually, it's a whack over the head with a shovel, or a .22 rifle that finishes the job. My dad doesn't remember which.

To collect the bounty, one presented the fox to the local RUC station, where the constable on the desk would, with a knife or a pair of scissors, cut off the tip of the fox's tongue. The constable would write out a receipt and pay you the bounty. By removing the tongue, the idea is that you could not, like a dame at a strange gala, visit station after station with a fox around your shoulders, collecting multiple bounties for the same creature. However, the constable in Hilltown, which was the local station, was young and either lazy or squeamish or both, and wouldn't do it. *Will you take it outside and do it yourselves?* he said to my dad and his brother Brian and Pat McGinn and Whiskey. The three of them said *yes*.

Naturally, they went on to Rathfriland, Katesbridge, and Loughbrickland where, by chance, they encountered one unwilling constable after another. At Loughbrickland, they called it quits, four times richer than they ought to have been. One fox for the price of four.

*

Thinking back to it, that visit before Christmas in 2019 was significant for several reasons. It was the first time in a while I'd heard my dad tell a story, which he did often when I was a child. It was also among the first times he'd spoken in any detail about what seemed to me the mysteries of his childhood and the subsequent traumas of adulthood in Northern Ireland in the 1970s and '80s; a reasonable disinclination.

More than anything, though, this story of fox hunting was so compelling because, in one of those wild happenstances around which stories coalesce, I'd been reading a slim Dover Children's Thrift Classics volume of *Japanese Fairy Tales*. I'd picked it up in the Oxfam bookshop on Ann Street in Belfast, and it has never since been far from my desk. Published in 1992, it is a new selection of five tales from *The Japanese Fairy Book*, which was originally compiled by Yei Theodora Ozaki and published by Archibald Constable & Co. in 1903. One story stands out more than the others: 'The Tongue-Cut Sparrow'.

In this version of the story, a cheerful, hardworking old man lives with his wife, a 'regular crosspatch'. To abate the man's loneliness – they have no children – he keeps a sparrow as a companion. After long days chopping wood, he comes home to the sparrow, talks to her, teaches her tricks. One day his wife, frustrated by how much time and food he is giving the sparrow, leaves out rice-paste with which she intends to starch clothes. The sparrow, thinking it was left out for her, eats the paste. The furious wife seizes the sparrow and, with a pair of scissors, cuts out her tongue. The bird flees into the woods.

The old man cries himself to sleep. The next day, he is determined to find the sparrow and searches through bamboo groves until he does, astonished to discover her tongue has grown back. The sparrow holds a lavish feast for the old man; her daughters perform a sparrow dance for him.

As a parting gift, the sparrow offers the man the choice of two baskets: a small wicker basket and a large wicker basket. Thinking modestly and of the long walk back, the man selects the smaller basket, which turns out to be full of gold and silver. Having explained to his wife what happened, she calls him an idiot for taking the smaller basket and goes back herself for the large one, which, she reasons, must be full of even more treasure. The sparrow, ever hospitable, says, by all means, take the large basket. In some versions of the story, goblins and demons leap out and scare the woman to death. In the version I read – for children – the old woman is only mildly spooked by a few sprites. She pledges to change her miserable ways and she and the old man live happily forever, spending wisely the sparrow's fortune.

Like many folktales, this one has a moral pulse: the importance of friendship and compassion; the consequences of greed. Japan's folktales abound with foxes and sparrows. According to Fanny Hagin Mayer, a researcher and translator who surveyed three thousand or so folktales recorded in the Niigata Prefecture, the fox, the monkey, the horse, and the rat appear most often among the quadrupeds.

While we associate the fox with cunning, the fox in Japanese folk and fairy tales is possessed of an incomparable wiliness. In these stories, the fox is a shapeshifter and trickster who transforms – utterly convincingly – into human form, such as the folktales (in Keigo Seki's taxonomy), 'Kongô-in and the Fox'; 'Tricked into Becoming a Priest'; 'The Acolyte and the Fox'; 'The Fox at Hijiyama'.

These stories happen in 'the olden days', 'where even foxes were honest', so it's not remarkable, then, that in 'How a Fox Returned a Kindness', the titular creature, remorseful for having eaten an old man's beans, should beg forgiveness. He can make a lot of money for the old man, he says and transforms into a pony, which the old man sells to a rich man. The pony bolts, returns, and retransforms into a fox. Thereafter, the fox becomes a tea kettle and is sold to a priest who likes tea, in which form he lingers until he's hung over the fire. In 'The Fox-Wife', a man comes back from the bathroom one night to find two identical wives where before he had one. He sends one away only to discover years later (and after two sons), he'd dismissed the wrong wife: '"The truth is that I was a fox," she said, and leaving her two children behind, she ran away crying'. A fox in these stories is never quite itself: it is always something else. It is, in its fields by night, in its earths by day, full of metaphoric potential.

Among birds, the sparrow appears frequently in folktales. Farmers, Mayer writes, make a show of scaring sparrows from their fields, where they are drawn to ripening rice. So often, these stories begin with a misdemeanour of hunger: a creature consumes, in its desperation, someone else's property. Then the repentance and the amends and the rewards.

In all versions of 'The Tongue-Cut Sparrow' I've read, the old man becomes wealthy. In one, since his wife is no more, he adopts a son to have around the house and to help him spend the money. As one turns over versions of the story and considers their moral and ethical instruction, it's hard not to wonder about the necessity of violence. I mean, does the sparrow have to have her tongue cut out for the old man to be given riches? She's clearly sensitive to the couple's fragile economy, she knows he's virtuous; why the whole rigmarole of plot? Why gift the old man gold and silver only after so much pain? The answer, I expect, is that it's not enough to have goodness rewarded: poor character and violence must be punished.

The gifts themselves are symbolic of character and conscience: you get what you deserve. You don't, however, know what you deserve until you open a gift and either treasure sparkles or a claw goes for your throat. We might always have done something better, given a second chance. Given a second chance, the old woman might not have cut out the sparrow's tongue. The tongue stands for language and agency, and, by extension, some idea of truth or witness. In *The Government of the Tongue* (1989), Heaney describes the organ as 'representing both a poet's personal gift of utterance and the common resources of language itself'. Its violation is injurious to both the individual and her society.

Across our literatures, it's women who most often have their tongues cut out: Philomela had hers cut out by her rapist, and the same story is told in *Titus Andronicus*. There are numerous others. In the many versions I've read of 'The Tongue-Cut Sparrow', it's only in the Dover Children's Thrift Classics version that the sparrow is female.

I tried to write this story about my dad three years ago, in verse, interweaving the two stories of the foxes and the sparrow; the Christmas presents whose contents are reflective of one's conscience. I fancied the fox's tongue was cut out so he might not be able to say (in his henhouse of paradise) the name of his killer. The poem didn't work, but it ended how this essay ends. Which is like this: after an hour of conversation, Ciaran left. I ran a few errands for my dad and got my things together. I'd be back on Christmas morning, I reassured him. Starting up the car, through the windshield and the kitchen window, I watched my dad slowly come over to the table and consider the gifts. Then he set them down unopened and turned off the light.

Opening a Gate

John Berger

The ceiling of the bedroom is painted a faded sky blue. There are two large rusty hooks screwed into the beams and from these, long ago, the farmer hung his smoked sausages and hams. This is the room in which I'm writing. Outside the window are old plum trees, the fruit now turning raven blue, and beyond them the nearest hill which forms the first step to the mountains.

Early this morning, when I was still in bed, a swallow flew in, circled the room, saw its error and flew out through the window past the plum trees to alight on the telephone wire. I relate this small incident because it seems to me to have something to do with Pentti Sammallahti's photographs. They too, like the swallow, are aberrant.

OPENING A GATE

I have had some of his photographs in the house now for two years. I often take them out of their folder to show to friends who pass. They usually gasp at first, and then peer closer, smiling. They look at the places shown for a longer time than is usual with a photograph. Perhaps they ask whether I know the photographer, Pentti Sammallahti, personally? Or they ask what part of Russia were they taken in? In what year? They never try to put their evident pleasure into words, for it is a secret one. They simply look closer and remember. What?

In each of the pictures there is at least one dog. That's clear and it might be no more than a gimmick. In fact the dogs offer a key for opening a door. No, a gate – for here everything is outside, outside and beyond.

I notice also in each photograph the special light, the light determined by the time of day or the season of the year. It is, invariably, the light in which figures hunt – for animals, forgotten names, a path leading home, a new day, sleep, the next lorry, spring. A light in which there is no permanence, a light of nothing longer than a glimpse. This too is a key to opening the gate.

The photos were taken with a panoramic camera, such as is normally used for making wide-section geological surveys. Here the wide-section is important, not, I think, for aesthetic reasons but, once again, for scientific, observational ones. A lens with a narrower focus would not have found what I now see, and so it would have remained invisible. What do I now see?

We live our daily lives in a constant exchange with the set of daily appearances surrounding us – often they are very familiar, sometimes they are unexpected and new, but always they confirm us in our lives. They do so even when they are threatening: the sight of a house burning, for example, or a man approaching us with a knife between his teeth, still reminds us (urgently) of our life and its importance. What we habitually see confirms us.

Yet it can happen, suddenly, unexpectedly, and most frequently in the half-light-of-glimpses, that we catch sight of another visible order which intersects with ours and has nothing to do with it.

The speed of a cinema film is 25 frames per second. God knows how many frames per second flicker past our daily perception. But it is as if, at the brief moments I'm talking about, suddenly and disconcertingly we see *between* two frames. We come upon a part of the visible which wasn't destined for us. Perhaps it was destined for night-birds, reindeer, ferrets, eels, whales ...

Our customary visible order is not the only one: it coexists with other orders. Stories of fairies, sprites, ogres were a human attempt to come to terms with this coexistence. Hunters are continually aware of it and so can read signs we do not see. Children feel it intuitively, because they have the habit of hiding behind things. There they discover the interstices between different sets of the visible.

Dogs, with their running legs, sharp noses, and developed memory for sounds, are the natural frontier experts of these interstices. Their eyes, whose message often confuses us for it is urgent and mute, are attuned both to the human order and to other visible orders. Perhaps this is why, on so many occasions and for different reasons, we train dogs as guides.

Probably it was a dog who led the great Finnish photographer to the moment and place for the taking of these pictures. In each one the human order, still in sight, is nevertheless no longer central and is slipping away. The interstices are open.

The result is unsettling: there is more solitude, more pain, more dereliction. At the same time, there is an expectancy which I have not experienced since childhood, since I talked to dogs, listened to their secrets, and kept them to myself.

PENTTI SAMMALLAHTI
SWAYAMBHUNATH, NEPAL (TWO DOGS FORMING A CIRCLE), 1994
COURTESY OF THE ARTIST

The Courage of Turtles

Edward Hoagland

Turtles are a kind of bird with the governor turned low. With the same attitude of removal, they cock a glance at what is going on, as if they need only to fly away. Until recently they were also a case of virtue rewarded, at least in the town where I grew up, because, being humble creatures, there were plenty of them. Even when we still had a few bobcats in the woods the local snapping turtles, growing up to forty pounds, were the largest carnivores. You would see them through the amber water, as big as greeny wash basins at the bottom of the pond, until they faded into the inscrutable mud as if they hadn't existed at all.

When I was ten I went to Dr. Green's Pond, a two-acre pond across the road. When I was twelve

I walked a mile or so to Taggart's Pond, which was lusher, had big water snakes and a waterfall; and shortly after that I was bicycling way up to the adventuresome vastness of Mud Pond, a lake-sized body of water in the reservoir system of a Connecticut city, possessed of cat-backed little islands and empty shacks and a forest of pines and hardwoods along the shore. Otters, foxes, and mink left their prints on the bank; there were pike and perch. As I got older, the estates and forgotten back lots in town were parcelled out and sold for nice prices, yet, though the woods had shrunk, it seemed that fewer people walked in the woods. The new residents didn't know how to find them. Eventually, exploring, they did find them, and it required some ingenuity and doubling around on my part to go for eight miles without meeting someone. I was grown by now, I lived in New York, and that's what I wanted to do on the occasional weekends when I came out.

Since Mud Pond contained drinking water I had felt confident nothing untoward would happen there. For a long while the developers stayed away, until the drought of the mid-1960s. This event, squeezing the edges in, convinced the local water company that the pond really wasn't a necessity as a catch basin, however; so they bulldozed a hole in the earthen dam, bulldozed the banks to fill in the bottom, and landscaped the flow of water that remained to wind like an English brook and provide a domestic view for the houses which were planned. Most of the painted turtles of Mud Pond, who had been inaccessible as they sunned on their rocks, wound up in boxes in boys' closets within a matter of days. Their footsteps in the dry leaves gave them away as they wandered forlornly. The snappers and the little musk turtles, neither of whom leave the water except once a year to lay their eggs, dug into the drying mud for another siege of hot weather, which they were accustomed to doing whenever the pond got low. But this time it was low for good; the mud baked over them and slowly entombed them. As for the ducks, I couldn't stroll in the woods and not feel guilty, because they were crouched beside every stagnant pothole, or were slinking between the bushes with their heads tucked into their shoulders so that I wouldn't see them. If they decided I had, they beat their way up through the screen of trees, striking their wings dangerously, and wheeled about with that headlong, magnificent velocity to locate another poor puddle.

I used to catch possums and black snakes as well as turtles, and I kept dogs and goats. Some summers I worked in a menagerie with the big personalities of the animal kingdom, like elephants and rhinoceroses. I was twenty before these enthusiasms began to wane, and it was then that I picked turtles as the particular animal I wanted to keep in touch with. I was allergic to fur, for one thing, and turtles need minimal care and not much in the way of quarters. They're personable beasts. They see the same colours we do and they seem to see just as well, as one discovers in trying to sneak up on them. In the laboratory they unravel the twists of a maze with the hot-blooded rapidity of a mammal. Though they can't run as fast as a rat, they improve on their errors just as quickly, pausing at each crossroads to look left and right. And they rock rhythmically in place, as we often do, although they are hatched from eggs, not the womb. (A common explanation psychologists give for our pleasure in rocking quietly is that it recapitulates our mother's heartbeat *in utero*.)

Snakes, by contrast, are dryly silent and priapic. They are smooth movers, legalistic, unblinking, and they afford the humour which the humourless do. But they make challenging captives; sometimes they don't eat for months on a point of order – if the light isn't right, for instance. Alligators are sticklers too. They're like war-horses, or German shepherds, and with their bar-shaped, vertical pupils adding emphasis, they have the *idée fixe* of eating, eating, even when they choose to refuse all food and stubbornly die. They delight in tossing a salamander up towards the sky and grabbing him in their long mouths as he comes down. They're so eager that they get the jitters, and they're too much of a proposition for a casual aquarium like mine. Frogs are depressingly defenceless: that moist, extensive back, with the bones almost sticking through. Hold a frog and you're holding its skeleton. Frogs' tasty legs are the staff of life to many animals – herons, raccoons, ribbon snakes – though they themselves are hard to feed. It's not an enviable role to be the staff of life, and after frogs you descend down the evolutionary ladder a big step to fish.

*

Turtles cough, burp, whistle, grunt and hiss, and produce social judgements. They put their heads together amicably enough, but then one drives the other back with the suddenness of two dogs who have been conversing in tones too low for an onlooker to hear. They pee in fear when they're first caught, but exercise both pluck and optimism in trying to escape, walking for hundreds of yards within the confines of their pen, carrying the weight of that cumbersome box on legs which are cruelly positioned for walking. They don't feel that the contest is unfair; they keep plugging, rolling like sailorly souls – a bobbing, infirm gait, a brave, sea-legged momentum – stopping occasionally to study the lay of the land. For me, anyway, they manage to contain the rest of the animal world. They can stretch out their necks like a giraffe, or loom underwater like an apocryphal hippo. They browse on lettuce thrown on the water like a cow moose which is partly submerged. They have a penguin's alertness, combined with a build like a brontosaurus when they rise up on tiptoe. Then they hunch and ponderously lunge like a grizzly going forward.

Baby turtles in a turtle bowl are a puzzle in geometrics. They're as decorative as pansy petals, but they are also self-directed building blocks, propping themselves on one another in different arrangements, before upending the tower. The timid individuals turn fearless, or vice versa. If one gets a bit arrogant he will push the others off the rock and afterwards climb down into the water and cling to the back of one of those he has bullied, tickling him with his hind feet until he bucks like a bronco. On the other hand, when this same milder-mannered fellow isn't exerting himself, he will stare right into the face of the sun for hours. What could be more lionlike? And he's at home in or out of the water and does lots of metaphysical tilting. He sinks and rises, with an infinity of levels to choose from; or, elongating himself, he climbs out on the land again to perambulate, sits boxed in his box, and finally slides back in the water, submerging into dreams.

I have five of these babies in a kidney-shaped bowl. The hatchling, who is a painted turtle, is not as large as the top joint of my thumb. He eats chicken gladly. Other foods he will attempt to eat but not with sufficient perseverance to succeed because he's so little. The yellow-bellied terrapin is probably a yearling, and he eats salad voraciously, but no meat, fish, or fowl. The Cumberland terrapin won't touch salad or chicken but eats fish and all of the meats except for bacon. The little snapper, with a black crenellated shell, feasts on any kind of meat, but rejects greens and fish. The fifth of the turtles is African. I acquired him only recently and don't know him well. A mottled brown, he unnerves the greener turtles, dragging their food off to his lairs. He doesn't seem to want to be green – he bites the algae off his shell, hanging meanwhile at daring, steep, head-first angles.

The snapper was a Ferdinand until I provided him with deeper water. Now he snaps at my pencil with his downturned and fearsome mouth, his swollen face like a napalm victim's. The Cumberland has an elliptical red mark on the side of his green-and-yellow head. He is benign by nature and ought to be as elegant as his scientific name (*Pseudemys scripta elegans*), except he has contracted a disease of the air bladder which has permanently inflated it; he floats high in the water at an undignified slant and can't go under. There may have been internal bleeding, too, because his carapace is stained along its ridge. Unfortunately, like flowers, baby turtles often die. Their mouths fill up with a white fungus and their lungs with pneumonia. Their organs clog up from the rust in the water, or diet troubles, and, like a dying man's, their eyes and heads become too prominent. Towards the end, the edge of the shell becomes flabby as felt and folds around them like a shroud.

While they live they're like puppies. Although they're vivacious, they would be a bore to be with all the time, so I also have an adult wood turtle about six inches long. Her top shell is the equal of any seashell for sculpturing, even a Cellini shell; it's like an old, dusty, richly engraved medallion dug out of a hillside. Her legs are salmon-orange bordered with black and protected by canted, heroic scales. Her plastron – the bottom shell – is splotched like a margay cat's coat, with black ocelli on a yellow background. It is convex to make room for the female organs inside, whereas a male's would be concave to help him fit tightly on top of her. Altogether, she exhibits every camouflage colour on her limbs and shells. She has a turtleneck neck, a tail like an elephant's, wise old pachydermous hind legs, and the face of a turkey – except that when I carry her she gazes at the passing ground with a hawk's eyes and mouth. Her feet fit to the fingers of my hand, one to each one, and she rides looking down. She can walk on the floor in perfect silence, but usually she lets her plastron knock portentously, like a footstep, so that she resembles some grand, concise, slow-moving id. But if an earthworm is presented, she jerks swiftly ahead, poises above it, and strikes like a mongoose, consuming it with wild vigour. Yet she will climb on my lap to eat bread or boiled eggs.

If put into a creek, she swims like a cutter, nosing forward to intercept a strange turtle and smell him. She drifts with the current to go downstream, manoeuvring behind a rock when she wants to take stock, or sinking to the nether levels, while bubbles float up. Getting out, choosing her path, she will proceed a distance and dig into a pile of humus, thrusting herself to the coolest layer at the bottom. The hole closes over her until it's as small as a mouse's hole. She's not as aquatic as a musk turtle, not quite as terrestrial as the box turtles in the same woods, but because of her versatility she's marvellous, she's everywhere. And though she breathes the way we breathe, with scarcely perceptible movements of her chest, sometimes instead she pumps her throat ruminatively, like a pipe smoker sucking and puffing. She waits and blinks, pumping her throat, turning her head, then sets off like a loping tiger in slow motion, hurdling the jungly lumber, the pea vine and twigs. She estimates angles so well that when she rides over the rocks, sliding down a drop-off with her rugged front legs extended, she has the grace of a rodeo mare.

But she's well off to be with me rather than at Mud Pond. The other turtles have fled – those that aren't baked into the bottom. Creeping up the brooks to sad, constricted marshes, burdened as they are with that box on their backs, they're walking into a setup where all their enemies move thirty times faster than they. It's like the nightmare most of us have whimpered through, where we are weighted down disastrously while trying to flee; fleeing our home ground, we try to run.

I've seen turtles in still worse straits. On Broadway, in New York, there is a penny arcade which used to sell baby terrapins that were scrawled with bon mots in enamel paint, such as KISS ME BABY. The manager turned out to be a wholesaler as well, and once I asked him whether he had any larger turtles to sell. He took me upstairs to a loft room devoted to the turtle business. There were desks for the paper work and a series of racks that held shallow tin bins atop one another, each with several hundred babies crawling around in it. He was a smudgy-complexioned, bespectacled, serious fellow and he did have a few adult terrapins, but I was going to school and wasn't actually planning to buy; I'd only wanted to see them. They were aquatic turtles, but here they went without water, presumably for weeks, lurching about in those dry bins like handicapped citizens, living on gumption. An easel where the artist worked stood in the middle of the floor. She had a palette and a clip attachment for fastening the babies in place. She wore a smock and a beret, and was homely, short, and eccentric-looking, with funny black hair, like some of the ladies who show their paintings in Washington Square in May. She had a cold, she was smoking, and her hand wasn't very steady, although she worked quickly enough. The smile that she produced for me would have looked giddy if she had been happier, or drunk. Of course the turtles' doom was sealed when she painted them, because their bodies inside would continue to grow but their shells would not. Gradually, invisibly, they would be crushed. Around us their bellies – two thousand belly shells – rubbed on the bins with a mournful, momentous hiss.

Somehow there were so many of them I didn't rescue one. Years later, however, I was walking on First Avenue when I noticed a basket of living turtles in front of a fish store. They were as dry as a heap of old bones in the sun; nevertheless, they were creeping over one another gimpily, doing their best to escape. I looked and was touched to discover that they appeared to be wood turtles, my favourites, so

I bought one. In my apartment I looked closer and realised that in fact this was a diamondback terrapin, which was bad news. Diamondbacks are tidewater turtles from brackish estuaries, and I had no seawater to keep him in. He spent his days thumping interminably against the baseboards, pushing for an opening through the wall. He drank thirstily but would not eat and had none of the hearty, accepting qualities of wood turtles. He was morose, paler in colour, sleeker and more Oriental in the carved ridges and rings that formed his shell. Though I felt sorry for him, finally I found his unrelenting presence exasperating. I carried him, struggling in a paper bag, across town to the Morton Street Pier on the Hudson River. It was August but grey and windy. He was very surprised when I tossed him in; for the first time in our association, I think, he was afraid. He looked afraid as he bobbed about on top of the water, looking up at me from ten feet below. Though we were both accustomed to his resistance and rigidity, seeing him still pitiful, I recognised that I must have done the wrong thing. At least the river was salty, but it was also bottomless; the waves were too rough for him, and the tide was coming in, bumping him against the pilings underneath the pier. Too late, I realised that he wouldn't be able to swim to a peaceful inlet in New Jersey, even if he could figure out which way to swim. But since, short of diving in after him, there was nothing I could do, I walked away.

What's New, Pussycat?

June Caldwell

I can never get away from cats. They are there in my dreams, practising manipulation, being prize weirdos. And in my front garden also, chewing and rubbing their hormone-packed chins on the twenty-one Nepeta (catnip) bushes, getting rightly stoned. I am mum to two: Cloudy (8), who I adopted after being shown a photograph when drunk at a Panti show – *High Heels in Low Places* – in Vicar Street in 2014, and Zarko (7), who belonged to a neighbour but moved in with us six years ago. The day Cloudy arrived, my mother and I were sick with nerves. My mother's cat Minnie died in 1948 after being shot in the eye with an air gun in Inchicore by a nasty teen neighbour, and I'd never had a cat growing up as my father didn't allow pets. We sat in the dining room on old brown tartan armchairs looking sideways at this tiny three-month-old, grey-and-white ball of fluff. Cloudy would hide under the large dining table on a velvet chair and every few minutes she'd lift the tablecloth with her paw to look at us and assess her situation. This went on for weeks. It was a year before she trusted us. For two years the neighbour who originally adopted Zarko from Cats Aid called to the door demanding him back. 'You can't legislate for where a cat chooses to go or live,' I said. They don't have a cat flap, we do. Case closed. His house had a few dogs and kids; Zarko wanted a quiet(er) life and foolishly chose our place instead.

Cats also psychically hunt me down as soon as I have the audacity to leave home for longer than a day. During a recent trip to a writer's retreat in the Loire Valley, I attempted to evade the forty-degree heat by visiting the Château d'Oiron, a sixteenth-century palace that is also a contemporary art gallery. The place is a mish-mash, with dozens of mid-sixteenth-century paintings among stucco plaster work showing the history of Troy and three episodes of the *Aeneid* by Virgil. There's also a wave tower with a sci-fi floating hemisphere, a whirring robot that interprets objects on the unreachable-to-some upper floors, as well as other curios by at least seventy artists. I learn that the original owners were the Gouffier family, whose grandson Claude was the real-life inspiration for Charles Perrault's *Puss in Boots*.

Claude Gouffier was an avid art collector who, in 1540, transformed Oiron from an ordinary castle to an incredible showy abode, creating an upper gallery for his own personal collection of oddities, a fifty-metre-long room to exhibit paintings and a round tower (where random items sit as part of a Marina Abramović performance piece in absentia). In 1551, King Henry II of France and his entire court were guests of Claude's, who was granted the title Marquis de Caravaz for playing ball. Charles Perrault loosely disguised lucky Caravaz as 'Marquis de Carabas' in his story of the lying cat who always gets what he wants. As you know, Puss in Boots was an anthropomorphic fellow who used trickery and deceit to gain power and wealth, and the prestige of the ruling classes.

Puss in Boots reminds me of modern-day 'pet-fluencers' who reel in the emotions of the masses while earning anything between $5,000 and $30,000 per Instagram post. These ordinary, non-earning cats do commonplace things like collecting leaves obsessively (see @theres_something_about_maryy). I adore them for their goofiness, performative prowess, individuality, and for being utterly uncontrollable and unpredictable. Once Instagram spots that I 'like' these posts, I'm sucked into the daily algorithms and slowly become addicted. Do I see myself in these cat behaviours? (Yes!) Do I desire the same nonsensical freedoms and unadulterated lack of responsibility? (Absolutely!) I can formally acknowledge that watching cats doing dumb shit online makes me feel more connected to something greater than the daily toll or the endless drudgery of human communication. Every cat owner I know feels the same.

#Catsofinstagram has in excess of 188 million posts and a net worth of $51.93 million. I end my visit to Château d'Oiron by sitting in a dark panelled lobby scrolling these new Puss in Boots on my phone. I've had enough of high-end art, so I switch from Instagram back to Twitter's @Bodegacats_ which features cats in small convenience stores doing their own thing: asleep on cereal packets, curled up on crates, pressed against fridge doors, peering out behind tins of beans.

Zarko and Cloudy don't get on, to put it mildly. He attacks her, is hideously jealous of her. She is disgusted by him, revolted even. She blames us for allowing him to live here. My partner and I have to keep both cats segregated 24/7. Zarko is locked in the sitting room at night, so Cloudy can come and go in peace from her back-garden hunts. During the day we lock her out until she appears at a window to be fed, so he can roam the house. We have a limited lifestyle because of this routine. My mother loved Cloudy but as she got older, the cat tired of her escalating frailty and now and then would hit her across the face. We wondered if Cloudy perceived old age as an evolutionary burden. By contrast, when my mother died in 2020, Zarko mourned her absence by refusing to roll on his back for seven months. Cats love to do this to show trust; how friendly and fun they are;

how much they enjoy your company. Instead, he made daily visits to her bedroom to scream at her pillow. (One of the most common signs of cat grief is increased yowling.) Cloudy didn't care. While she's nocturnally affectionate, licking my chin or forehead during the witching hours – once I woke up simultaneously fainting when she was doing this – she is blatantly disinterested in people and especially children.

Growing up in Ballymun in the seventies, no one cared about cats. It was dogs all the way, followed by *Top of the Pops*, cardboard packets of reconstituted food, and fear of priests. There was just one cat on the road, Fairy Forde. His proprietor was a feminist (instrumental in maternity leave being introduced into Ireland) and some of the neighbours quipped about her needing a cat, i.e. being a witch. That long-held injurious association. Women + cat(s) = witch = unknown powers = ungodly. Fairy was a robust marmalade who spent his days watching us play, but never played *with* us. He had zero interest. Cats can't be bothered so much of the time. They lack the cognitive skills to interpret human language, but they recognise when you talk to them and can also recognise the sound of their name. Cat behaviourists believe an adult feline's intelligence is comparable to that of a two-year-old human toddler.

Cats stare. Cats stare at us. Cats stare at us and it makes us feel vulnerable. It unsettles us. Their eyes glow due to a mirror-like structure behind their retinas that enables them to have night vision. Cats stare at us at night and their eyes glow and it scares us because we cannot see in the dark. Or as one *Washington Post* journalist put it in yet another cat article in time for yet another Hallowe'en: 'The glow is not the most unsettling thing about cats' eyes. Unlike tigers and jaguars and other big cats, house cats have vertically slit pupils, a common feature among small nocturnal predators that hunt close to the ground. What else has vertically slit pupils and also occasionally hisses? The serpent. And who made his first biblical appearance as a snake? That's right: Satan.'

In school we were told that the Celts believed cats were human souls forced to return to Earth after committing bad deeds. It fascinated me to learn that Egyptians considered them divine (in 1888 scientists found a cat cemetery in Beni Hassan brimming with tens of thousands of cat mummies, though they were sold off as fertiliser by the ton), and killing a cat was forbidden. The Greek historian Herodotus reported that whenever a household cat died, the entire family had to mourn and shave their eyebrows.

These days, as a vegetarian and someone who believes it's time we reached a level of psychological sophistication where we stop eating animals for protein, I find it too upsetting to list the places around the world where cat meat is still enjoyed or used in remedies (treatment of arthritis or as an aphrodisiac). Dog fighters use markers to colour the white parts of cats and kittens so they can bet on which colour will die first. I also find it very strange that if you crash into a dog while driving you must report it to the Gardaí, but there is no onus on reporting the same incident if it involves a cat. I prefer instead to concentrate on more palatable facts: how cats are disjointed all over with bendy spines that squeeze into the smallest of spaces; how they activate 100 percent of their muscles when running (up to 48km/h); jump up to five times their height; the little hairs in their ears ('ear furnishings') reorientating their 'righting reflex', so they land on their feet; how their demanding purrs can reach a pitch of 520-hertz to mimic a baby's cry. It makes me think of Rossini's *The Cat Duet*, a popular performance piece for two sopranos, with lyrics that consist entirely of the repeated word *'miau'* ('meow').

My cat Cloudy sleeps on a linen-clad, feather-filled cushion that Nuala O'Faolain had on her writing chair in Dublin when she wrote *Are You Somebody?* – and later in her cottage in Co. Clare. Her family gifted me the cushion after I curated an exhibition on her memoir for the Museum of Literature Ireland (MoLI) in 2019. Nuala loved cats. I interviewed her once while doing a journalism postgrad and two weeks later she called to my bedsit with a cat in a basket, and an envelope with £30 to procure 'Sandra' a hysterectomy. The cat was a gift, allegedly. I was a broke student so spent the money on food and booze, and the cat got pregnant weeks later, something I hid from Nuala for years.

The other cat Zarko has recently been diagnosed with congenital hip dysplasia, a condition I've also had since birth. We pay €55 per month for a new-fangled nerve blocker Solensia that targets nerve growth factor (NGF), a key driver of osteoarthritis pain. I find myself feeling an odd jealousy as there's no human form of this injection yet, and I am lumbered with the same pain we've just rid him of. He cries all the way to the aptly named Fox Veterinary Clinic in Finglas, and all the way back again. He doesn't understand why we would cause him such distress.

Every Hallowe'en I wonder about the malevolent black cat thing, where it really comes from? It's a well-worn narrative by now but allegedly in 1233, an official church document called 'Vox in Rama' was issued by Pope Gregory IX, condemning black cats as an incarnation of Satan. This resulted in the mass murder of cats throughout Europe, allegedly, giving social historians a handy follow-on: there weren't enough cats around, as a result, to kill the overspill of rodents, especially diseased ones carrying the bubonic plague. If this is true, it's fiendishly karmic. Cats were always 'worked' to keep grain supplies free of gnawers, operating alongside women and implements such as brooms and brushes to ensure homesteads, shops, farms, stayed clean. The association with the cats and the devil quickly moved on to women's sexuality and strange lustful powers, of course.

In the *Malleus Maleficarum*, the medieval treatise on witchcraft, a thirteenth-century folk story is recounted: three witches turn themselves into cats, attack a man on the street, and accuse him of assault in court, showing scratches on their bodies. From then on, witches were believed to have cats as familiars, or to morph into felines at night themselves. This 'dark' and devilish association continues through the centuries in literature and art. Manet's *Olympia*, painted in 1863, depicts a nude woman with her hand over her vagina, a small black cat shrouded in darkness at her feet. The cat is somehow uncomfortable to look at, wicked, malicious. Manet got a lot of stick for it. Years later he painted a more wholesome work of his wife with a black-and-white cat sitting cosily on her lap: *Woman with a Cat*. It didn't bury the controversy. The wide-grinning Cheshire cat in *Alice's Adventures in Wonderland* was no less creepy in book form two years after *Olympia*: 'The Cat only grinned when it saw Alice. It looked good-natured, she thought: still it had very long claws and a great many teeth, so she felt that it ought to be treated with respect.'

On a press trip to Bath to review a play for the *Sunday Business Post* in 2014, I had a strange dream that confirms the mystical strangeness of cats firsthand. Cloudy was summoning me home, using *Macbeth*'s witches to get her message across. I woke up in a sweat but it somehow made sense. The hags from *Macbeth* didn't plunge their 'brinded' cat in the cauldron with the newt's eye and dog tongue – they kept it alive and meowing for a reason. I rang home and my mother confirmed Cloudy was missing. She was six months old and managed to climb over the back-garden wall for the first time and got lost. When I returned two days later and called out for her, she responded with a long weeping sound that she uses only in my company. She was hiding under a bush in a neighbour's garden a few doors down, tired and dehydrated. This was the first telekinetic episode between us, with many more since. She uses this same telepathy to tell me she's at the window in the sitting room at night and wants in. Zarko is far more grounded; he waterwheels the glass so you can hear his needs. Cloudy messes with the mind.

The 1961 Acoustic Kitty Project in America is the perfect example of how people, even highly skilled and intelligent people, underestimate the innate behavioural characteristics of cats. The project used a tech-upgraded 'FrankenKitty' as a real-life spy, at the cost of $20 million. The CIA had already attempted something similar with ravens, though it failed. It took five years to plan; exploiting a cat's innate curiosity and the ability to come and go unnoticed. With the use of audio cues, the cat would do as it was trained to do and go where it was supposed to go, recording conversations useful to the government. A cat's tail was used as antenna, wires mangled all the way up the spine, with a microphone in the ear canal, while a battery was lodged in the chest area. In the field on a first test-run, handlers watched on as the cat was killed by a taxi. WikiLeaks documents show research conclusions that any pet owner knows instinctively: you can only train cats to move short distances. 'The environmental and security factors in using this technique in a real foreign situation force us to conclude that for our (intelligence) purposes, it would not be practical,' a memo in the WikiLeaks arsenal reads. How the cat didn't reef it off her or pull at the many surgical sites is a miracle. Operatives gathered the cat's remains to stop the Soviets from retrieving expensive audio equipment and the entire project was abandoned in 1967. Following the abject failure of the Acoustic Kitty Project, the CIA moved on to exploring psychic phenomena in military and domestic intelligence applications in what became known as the Stargate Project. This later part-inspired the hit Netflix series *Stranger Things* but the cat-astrophe that was Acoustic Kitty has so far only made it into a ten-minute short.

Zarko is fat, so much so that everyone who visits our house says so. 'That cat is huge!' He's not an active cat, like Cloudy. He doesn't move much and hangs out in the house most of the day. He's never caught or killed a bird; he chases butterflies if he makes it out to the back garden at all. The most evil thing he's ever done is eat spiders. But his size and girth continue to cause us grief, and him pain. Sometime last year a gigantic 3D cat appeared on a billboard in Tokyo, and I was delighted to see a digital cat allowed to be enormous and loved, even funky. It loomed mischievously over a busy train station, moving about on a 1,664-square-foot curved LED screen. The 4K-resolution display showed the cat walking around high above the Japanese capital as it audibly meows. We've been in the company of domestic cats for ten thousand years. They will be there with us when we are part-machines in the far-off future. I like to think they've already mastered strange new pathways of physics and are waiting for us, sprawled, Zarko-like, on free-floating velvet couches, with translator voice boxes to tell us clearly what they expect of us and how exactly we should deliver.

Precious

Jessica Traynor

When I was growing up, neighbours around ten houses away had a Weimaraner. It was sculptural in its perfection. It bit every member of their family. It was called Taupe.

Most dogs I knew growing up were called Brandy or Guinness or Whiskey, but the family who owned Taupe had white-painted walls and art in thin frames, and so instead of calling their dog after a drink, they called him after an intangible shade that existed somewhere north of beige, south of mushroom. It was a ghostly shade, and Taupe was a ghostly dog, appearing in the corner of your eye, not making any noise at all until he was close enough to begin barking and slavering.

Once when I was alone in the lane that ran behind our houses, Taupe escaped his back garden and came at me, growling. I ran. In my mind I can feel Taupe's breath on the back of my neck, the singing of the nerves in my shoulders as his jaws inched closer. But of course, this is most likely a memory constructed later, one that coagulated around the fear of dogs that stemmed from whatever the truth of this incident was. It feels very real though. My mother remembers me slamming the back door and standing in the kitchen gasping *TAUPE TAUPE TAUPE*, which must have been confusing at the time. The dog bit the family's baby soon after, and was put down. Their next dog, a Maltese, was stolen from their back garden a few months later. I'm not sure whether they had bad luck with dogs, or dogs had bad luck with them.

For a long time I've been interested in the idea of the uncanny; those things that strike you as initially familiar, before revealing their strangeness in a manner that makes you feel like something sacred and safe has been violated. Sigmund Freud, in his 1919 essay 'The Uncanny', talks about the transformation of the idea of what is *heimlich* or 'homely' to *unheimlich* – which does not necessarily mean 'unhomely', but rather 'a hidden, familiar thing that has undergone repression and then emerged from it'. It's the very familiarity of the uncanny that provokes our fear. For a long time after the Taupe incident, I was afraid of dogs. This encounter is my earliest clear memory of being afraid, and this fear was planted among my foundational memories of home, forever linking the uncanny, in my mind, to dogs – those least uncanny-seeming of creatures.

Because of this fear, we had cats, and I became attuned to their ways of being, which are very different to dogs' ways of being. I remember, as a child, the process of learning how to exist in the presence of cats; how to adapt my movements to make them comfortable, in a way that's such second nature to me now that when a person unused to cats blunders into a room and startles a cat, I feel affronted on the cat's behalf. My dad was an inveterate cat-startler, always sweeping noisily into rooms. One day he opened the door to our kitchen and was brought up short by the sudden stillness of the five cats who had positioned themselves around the room and who up until his entry had been engaged in various acts of grooming or lolling about. My dad started to walk carefully through the kitchen. I watched from the other end of the room. He got about a third of the way across the floor before the nearest cat to him made a break for the back door, sending the others whirling about the room in a hurricane of shed fur, their claws shredding soft furnishings as they went. *For JAYSUS sake*, my dad erupted, *I'm sick of fuckin' pussyfootin' around my own house!*

I felt the same way about dogs for a long time. I wanted to understand them, but there was that fear – even their effusiveness alarmed me. If confronted with one that seemed friendly, I would try to pet their heads and they would try to lick my hand and I would try again to pet their heads like that 'One potato, two potato' game where you stack hands endlessly on top of each other, but no one gets petted and no one's hand gets licked. They would jump and bark and I never understood why.

Then one Christmas our neighbour, a man who was caring for his dying mother, got a puppy he called Precious. Precious was a small black bullet-shaped mongrel terrier with an empathetic gaze. Our neighbour brought her to our house on Christmas day where she full-body wriggled around the room and pissed on the carpet.

Precious must have had some springer in her, because no wall could confine her. She would regularly turn up in our back garden uninvited, staring hopefully through the glass in our back door. Learning her movements and boundaries diminished my fear of other dogs to a manageable level. Eventually, I would go to my back garden and call her and she would bound over the walls between the gardens. Even the cats had a lot of time for her, all things considered, and I'd set off into the back lane with the dog and most of the cats in tow. Sometimes our neighbour would let me take her for walks to the local park, where she would exhaust herself straining on her lead. But I got older, and pets became less central to my day-to-day existence. When I was thirteen, my parents' marriage ended and we moved away from the area. Precious continued to exist within the bubble of my childhood, which never escaped the walls that she had leaped with such ease.

In an episode in Claire Kilroy's 2012 novel, *The Devil I Know*, Tristram St. Lawrence, our amoral protagonist, is exploring the grounds of his ancestral home with avaricious property developer Hickey, who wants to build on the land. The pair make an unnerving discovery. A childhood pet of Tristram's, a pony called Prince, has been abandoned in the grounds and re-emerges from the undergrowth. 'The damage', as Tristram refers to the pony, has come back, bursting from the unconscious bringing with him connotations of a childhood innocence not lost, necessarily, but decayed. It's an immensely troubling scene, a moment of deep uncanniness. *How could the pony still be alive?* I found myself asking while reading, *this makes no sense, surely someone would have done something, surely it couldn't have survived so long…* The reader shares Tristram's horror at this confrontation, and his urge to deny its reality. But there, undeniably, is the pony. Tristram says of him:

> Soon the girls were no longer girls but he wasn't to know that, and so he'd been waiting for them to return ever since, wondering, if ponies can wonder and I fear they can – I fear that every blessed thing on this earth is cursed with the capacity to wonder at its predicament – Prince was left wondering what he'd done wrong.

Tristram runs away from this unsettling confrontation and does what he always does – uses his money to make 'the damage' disappear by having the animal put down. There's something in this episode in the book that I find uncomfortably moving. Is it anthropomorphism? Yes of course. But there's something in the image of that pony that seems to represent innocence doomed to suffer. It's a cruel dynamic, but it exists, no matter how I try to ignore it.

When I was in my early twenties, long after we'd moved away from my childhood home, a surprise caller came to my mother's door. It was late spring and I was upstairs studying for college exams. A voice that tugged at the edges of memory pulled me downstairs, where our old neighbour, Precious's owner, was standing in our doorway, talking to my mother. He looked older, a little frailer. For a moment I saw the shadow of his mother in him, a wraith-like figure from my childhood, who would throw coins out the window and shout at me, 'Boy! Get me some cigarettes!' Our neighbour was chatting to my mother – his house had burned down, and by astonishing coincidence, he had rented a house on our road while his own was being renovated. I stood on the stairs (the hallway was tiny) and we all made small talk. It took me a moment to notice the lead in his hand. Eventually our neighbour moved his bulk from the doorway and we saw her. She was crouched on the path behind him, black fur gone grey. Even from that distance I could pick out the tracery of her ribs. Her eyes were milky, and she turned her head only slightly when I said her name.

'Precious?'

It couldn't be her, I thought. She had been old when my parents' marriage had ended, old when, seven years previously, we had left that neighbourhood behind us. This shivering thing had to be another dog. The misery that gripped me at the sight of her felt a lot like fear. Freud, quoting Jentsch in 'The Uncanny' says, 'In telling a story, one of the most successful devices for easily creating uncanny effects is to leave the reader in uncertainty whether a particular figure in the story is a human being or an automaton'; here, as in *The Devil I Know*, the subject that confronts us is not an automaton, but a revenant. The effect is the same.

Our neighbour confirmed it was Precious, said something about her joints, said something about her sight, but I wasn't really listening. He pulled her lead until she crept into the house and sat in our living room, hunched, not knowing where she was. She looked distressed. I put my hand on her grey muzzle hoping for some flicker of recognition. But there was nothing.

I made my excuses and went back upstairs to my room to recover from the shock, from the sense that something in the natural order had been disturbed. I sat on my single bed and listened to the voices rising and falling downstairs. I listened for the pauses in the conversation where I might hear the dog's shuddering wheeze. In the literature for Mike Kelley's 2004 exhibition, *The Uncanny*, at Tate Liverpool, the idea of the 'double' is invoked – 'the disturbingly realistic representation of the human figure suspended between life and death'. Precious, it seemed to me, was trapped in that in-between state: a place without any sense of finality, or resolution, or potential.

I sat on the edge of the bed staring out the window for a long time after the door closed and Precious retreated back into the strange limbo she'd crawled out from. We all know on some level that death is the ultimate kindness, and the withholding of it the stuff of nightmares. For a long time I've wondered if the emotion I experience when I recall this memory is displaced sadness about my parents' divorce; the loss of our family home, that premature closing of the door into childhood. To be honest, none of these things move me the way the memory of that dog does.

River before Me

Tim MacGabhann

1 Horse Dreams

Fair share of run-ins with horses / one pair one night in Buenos Aires / key drooping from the chain around my neck / in a bar that creaked like a haunted galleon / necks of the two lads from whom I'd bought the crack / lengthening slowly into necks of a chestnut mare and stallion / heads jerking up and down / teeth nibbling lips / teeth white with froth / and a cough working hard up and down in one of my lungs / like that wound fishing line people use to catch pike /

And / *liftoff* / My skin shimmered / and evaporated / heat mirage above a runway / My tendons frayed /

broke with a ping / my jaw yawned itself off, dropped into fine / rust-coloured sand with the rest of my bones /

So when I woke this morning with a sand-coloured gelding / tethered to my bed / and the fizz of a nightmare / pint still spritzing my cheek / I thought I was back / to my old using days / Horse whickered / shook his mane / laughing off my fear / almost like / then lowered his eyes to mine / The flipped blur of what he saw showed no future / worth being afraid of / Slowly that electric / prickle of sweat left my feet / Next time I awoke it was to a yellow room loud with sun / the bell-clang of a machete's slow trochees / a hose outside unspooling the same clean / white 'S' over / and over again /

Through the window I watched a cricket / pick her way over the telephone wire / joining the neighbour's wall to ours / then burrow out of view / into a safe / dry crack /

2   Cormorants

I

Basement apartment / Rathmines or Rathgar / took one suck too many on too strong of a joint / saw host morph every second / second into a cartoon / polar bear chuckling on his ice-floe / me I tumbled / arse over / tip / through billows of sand / through a seafloor / that could have been / yellow smoke / Next day / more than / before / could smell a lightbulb / deep in my brain / filament burning out with a yellow whine / a mosquito's / smell of it / yellow smoke / Half an acid tab / turned the sirens / wailing down Pearse Street / into arias / *Tosca* / *Turandot* / even Verdi's / *Requiem* /

II

Awake and aphasiac for days / burnings I thought of as epiphanies / moving and turning in my head / navy cloth heart / hung on a wall / stuck through with seven tin swords / old man flying over sad / sooty / cities / carried by a broken umbrella / hair and beard / like old Brillo pads / black plastic hanging tattered / in three 'Y'-shapes / from the struts of an empty ad / hoarding / Dismas Gestas the other fella / poignant yellow lights of boats way out on the sea / air pollution's orange clouds / fulgurating / between the chimbleys / on the strand sacking / frayed ropes / toppled stone urns / rusted sheet metal / brown fog / one long / sweep / into release

III

Greed less for enlightenment / than for / exhaustion / gallop laps of town / till tired enough to collapse / into sleep never as deep needed to be / Wired / turn a corner / see a cowled skull hovering before me / Spine turned to / Water / pasted myself to the wall / and the bone grin / resolved into the 'P' / of a parking meter /

IV

Clifftop at night / white zeroes spread / breaking / across loud water / white zeroes / calcites in the rock / all those / million-year-ago crashings and rendings / same sound / same instant / same / transience / ran finger along / stippling / long-gone cluster of minnow / massed coral / muscle fibre / shapes / a whole ocean / summarised / a scribble / lichen was / eating / nursed sternum with / knuckles / front / back / Wind tickled / scarf's tassels / across / cheek / Shake a million of me out of any century / fill your Barbusses / your Gides / your Dostoyevskys / their shite / acolytes / Oh so flip / and / inert till some invasion turns me / meek / fail to beg or / bargain / my way out of the infantry / trip in a barrel / day one / get shot in the neck /

V

Cormorants driven mad by light pollution / were diving from tall red diodes / in the tech quarter / Salt breaths cut in and out / scald as deep / as the belts of whiskey / with which I toasted each bird's dive / A poet so stricken by grief that he jumped off a cliff / was the first of these birds / Ovid said / A plume of foam rose lovingly along his body / as to smooth his torso / into a long sleek neck / But his wish to die survived this / alteration / That's why they dive down so far / rise with a yawk / of frustration / every goddam time / water in little no-worth pearls / from their flared wings / Patron saints of burnout / eyes blank / shoulders braced / unable to stay still /

VI

Proust / wet-eyed / ashen / scorched cardigan / velvet gloves / walked the silver lunar / ruin of Paris during the bombardment / nowhere of spotlights / criss-crossing / clouds / shell-whine / barrage balloons' mesh nets / like a portcullis in the air / plucked from his hatbrim long shards of shrapnel / saying to Céleste / *I can hardly see. So I found it beautiful* / Stooped like a cormorant / lilac insomnia / pouches under each eye / four pieces of furniture / to his name / green banker's lamp / drawn curtains / nobody else's books but his own / lilac cover / teeming ant-size / letters /

VIII

Head down / shoulders hunched / unable to come down from the pills / led by Proust's ghost past Boland's Mills / crumbling and golden there on the water / *LIVE DUBLIN DIE YOUNG* scrawled on one wall / tall red letters / Couldn't keep laugh / from kiting up out of me / brilliant red flags / hurt the throat / tugged up the floor of / stomach / like the spongey tautness / of a boxing ring / *What?* said Proust's ghost / *That*, I said / and waved / a hand at the graffiti / I put that in a poem / ever / no cunt will believe it / My voice / half ravé / half groan / too fucked to keep going / too fucked to / stop / Proust's ghost had disappeared / Eased myself onto a bench / electric scald of Huzzar / flowing down my throat / like melted / stars /

3 Dogs

I

Next door's dogs went / mental / all night / Barks chewed *Jungle Rain* setting / of my white noise machine / to ragged grey lace / hauled it up the road / cunts / Timed my breath to the blinking of a satellite / above the house / slept / dreamt of a bag of mince / about to burst / woke / bolted to the window / unbolted window / saw red wet ribbons of meat dragged over the road into town / Well / I know an omen when I / smell one / texted round / made sure nobody'd OD'd / recalled a / legend / Greek priest / fever / dreams / snakes twining on his chest / Woke up / cured / a dog / lapping his forehead /

II

Indoors / ill / Greek tomb steles / on the laptop / cypresses' lit black candles / hard blue light / like the first morning of the world /

One tomb showed a wolf / another / a man / crawling forward carrying an eggshell / A seated woman met her younger self / A fourth / chin in hands / hollows for pupils / sat on a cliff to watch the / sinking / of a ship's sail /

III

All day today / wind rose and fell / bellows under it / Laurel shreds and twigs coated the porch / Garden's angriest / thrush / on patrol /

Fig-dark circles under each eye / thinking of poems as test tubes / to hold grit left when thought burns out / Dust coats clear inner curve / pebbly nubs / shuffle with a lightweight anthracite clink /

Now mill / further / leave for / rain / black muck brimming / fresh ink / overflowing in big glossy slicks / to form some unutterable consonant / morass of sound to choke on / drown in / writing minus writer / minus / reader / O for that run-off of speech to leak across / flat rooftop / lap my bare toes / climb me / vitreous calm / coolness / detach from speech / float in a zone / of pure touch / dig into silence as into a beach / down to the wet stratum that will shake / and gleam like fat if you reach in / give it a smack /

Close now / Blue dark / white dust / cobbles' / bossy skulls under my heels / deep tunnel feel / to how the laurels cross overhead / Pine needles / Web of amate roots / has swallowed a boulder / split it slowly open / grunt of release / cool dark air sighed / out with a cellar odour / almost of apples /

Wind creases grass beyond tumbled wall / Blackbird cheeps / Two horses / wave their tails / I envy the dead / their calm / so total / so bluely / neutral /

IV

A road / Evening / White dust / thin manes / An arch with stone horses / red brick / worn plaster like bone / and muscle in cross-section / Statues blotched with lichen / beyond the gate / nearest a reclining woman / upturned torch in one hand / mirror in the other / shaded by mulberries and cypresses / near a colonnade / Under her lie / empty mask / split pomegranate / wilted poppies / Barely bright / Still bright / Walk back / River before me / Family of horses drinking / water's / shivering gold / Pat mare's flank / Feel scar / branded with / scorpion / two foals splash / circle / light combs glossily along / ribs / bodies steam / heat reaches / skin /

4 The Silence

I

Long time off the powder / my problems are earthenware words / *respect* / *caring* / I watch snooker / cool my head with the legend / of how / so he'd never have to sing again / Orpheus's last known prayer was to be turned into a swan / Apollo obliged / Now he floats over / the unruffled surface of tarns / listens to wind whine through his feathers / a white gap in the sky's big nothing / Not a fate he deserves / Ask any lover / twisted round the bend by his riffs / who'd prefer to see him bang / into a window / die without a scar /

II

Whole thing staged anyway / Ghosts whispered in Eurydice's ear / told her to stumble / utter a small cry / make Orpheus turn / so they could pry him apart / arm from shoulder / knee from socket / the lot chucked into a river / that washed his moaning head to / *an isle* / *bare of all but wind-creased grass and asters* /

Rewind the tape again / Orpheus hears a rustle / gives chase / all curious / over-earnest wonder / in the sun-dappled darkness / till the umbilical / lasso of whatever he'd rather not hear / lands round his neck / His whole job's to muster / in that instant / a trapezist's pretend-froideur / as his gaze fades out / on asphodels /

III

For years / I believed cruelty a form of knowledge / had myself skinned from chin to knee / heart liver lights melts / under a vitrine / resting on that plastic grass / wedge / of lemon nestled for freshness in the coils / of my entrails / food dye injected daily / to keep my heart a nice tint of aubergine / Trouble was nobody was buying / They'd sliver off a bit / dab with chipotle / mosey off with a noise of approval / to hide non-intention to return / Nights / alone / under flytrap's / blue shine / and hum / I'd finger my throat's inner fibres / picture piano wire / cotton-swaddled / hammers /

IV

Maybe I was never very human / To lean my forearms on cool / grainy teak / under a perfect cone of light and smoke / in a house by a port was my one ambition / But even to say 'I' at all keeps me beholden / to all that booj nonsense of lyricism / and its anxious ache after self-dissolution / as modulated out of a yen for cataclysm / ego-orgasm / warm leak of chrism flowing / down the forehead of some high-on-life enlightened / dickhead / who still thinks the laurel crown / still / exists / first off / and is desirable / second / self outside self / fingering cool mist / *Rückenfigur* fuckery / except with narcotics / and then / brutally / without / Liszt / that fraud / hammering the keys / till his outline / resolves / into *long combed streams of golden light* /

V

No / in the evening of life / if I make it / I'll be judged by whatever blew the valves / off my ego / and so shook my skinny frame / as to leave me standing there fully convinced / that I've stopped measuring my love / for things / by how far I might be lifted / above / supermarket vents and heat shimmer / and the rest of the slo-mo ruin cinema / playing forever behind my eyes / vivid ivies / their green seethe over rubble and balconies /

VI

Oh / yes / long time off the powder / but understand / the death I will always want will always have / smoke-tang / self / a cough / from a body / falling back / under a palm tree / light splitting fronds / into a crown of nine perfect white prongs / you've seen this ad I'm sure / head's final / transmission /

dying to a / burble / not even I'm arsed / hearing / never have been / all along /

Give me / consoling paisley fractals / that unfurl / from my pipe's / scorched glass bowl / Give me / crystal lattice / taste of whatever illegal ethyl / it is / holds crack rocks / together / crack open / self's / tomb-rock / Let / diagonals of hard light / spill in / like a big chord on a church organ /

Then, Horses

Sabrina Mandanici

On a map, the Ardèche looks like a womb. Driving through this mountainous region in southeastern France, the womb grows furrowed with rivers and streams. At its bottom, where the Ardèche gorges begin, lies the Pont d'Arc. With its towering, slightly slanted arch, this natural bridge resembles an enormous mammoth trudging through the river: a rocky gatekeeper sculpted by water. On either bank the mammoth is faced with forested limestone cliffs. If it stood on its hind legs, stretching its trunk to explore the cliffs' cavities, it would find the opening to the Chauvet cave, home to some of the oldest paintings in the world. I crossed their path by coincidence.

THEN, HORSES

One evening in late September, I took the night train to Paris, followed by a midday connection to Vézelay, the first stop on my medieval-church tour through Burgundy. I am not religious, but I've always felt drawn to French cathedrals. To their creature-filled portals and botanical adornments, the mystery of their stories carved in stone.

The conductor woke me at a small station in the middle of nowhere. He was asking everyone to leave the train until some 'technical problem' could be resolved. Three hours later, the train was declared irreparable. It was raining and there was no place to go. I stood under the station's canopy, watching cars pulling in and out from the parking lot, and sat on the steps to pull out my map.

'Do you need a lift?' a man called through the open window of an old Saab, his French shivering with an American cadence. He got out of the car and stepped under the canopy; his blond hair tousled by the wind. On my map he pointed out that Vézelay was on his way to Vallon-Pont-d'Arc. He knew the church well. 'The A6 runs right by, see?'

Anton was a palaeobiologist, specialising in large mammals from the Ice Age. He was on his annual trip to the Chauvet cave, where, some 30,000 years ago, the Cro-Magnons had painted and carved the most spectacular animals on and into the rock. After the collapse of a ceiling, the cave had been sealed for 20,000 years; it was rediscovered, by accident, in 1994. Before the Cro-Magnons arrived and after they had left, the cave was inhabited by bears, whose bones Anton was examining. He was part of a handpicked research team, the only people now admitted into the actual cave, while a replica had been built for the public. They met twice a year, for a fortnight, in spring and autumn. This year's session was cut short, however, because the cave's climate was unstable. Anton planned to use the remaining time to revisit other caves in neighbouring Dordogne. A friend (a zoologist, who wasn't part of the team), was supposed to join him, but had cancelled last minute. To make sure he hadn't changed his mind, Anton drove to the station nonetheless, and waited.

'Why Vézelay first?' he asked, with genuine curiosity.

'Because you have to start somewhere.'

I told him I'd recently left my first job, working for an art collector in Berlin, a job that, when I got it, after a sequence of unpaid internships, had seemed so significant; and then, four months in, I quit. Using the small savings I had, I decided to travel. I wanted to go back in time, to a place that made 'sense', although I did not know what 'sense' actually meant.

'Well why not start with the animals, before you go see the saints?' His friends were expecting two guests anyway, Anton said. There was plenty of room and they'd appreciate the company. 'And the food is going to be terrific,' he added, smiling.

Martine and Jean-Michel took me in as if I was meant to be there. An archaeologist and history teacher, both retired, their old farmhouse was filled with books, shards, and other sundry treasures brought back from their trips; each object had its own story, which Jean-Michel was eager to tell, and I was happy to listen. In the evenings we would gather in the kitchen, chopping vegetables and unwrapping cheeses, which the couple bought from a close-by farmer. They were as friendly with him as with his cows and goats, to whom they jokingly referred as the children they never had. 'This one is from Babette and Hirondelle,' Martine said placing the small chèvre on the board. They knew all the animals' names.

On our second morning, Anton and I headed for Dordogne, where we were planning to visit three caves in close proximity. Dawn was hours away, but the drive was long, and Anton wanted to reach the first cave by the time it opened at ten. 'Why mammals from the Ice Age?' I asked him, sipping coffee from a thermos cup, trying to stay awake. 'Because they are our ancestors,' he said, shifting gears. 'We owe it to them to tell their stories, no matter how fragmented they are.'

We arrived early at Rouffignac. The parking lot was still empty, the entrance locked behind a massive sliding gate. To shorten the wait we explored the small paths veering off the main road. There were benches and picnic areas, and occasionally the outlines of mammoths, life-sized and stenciled in steel, popped up between the trees. They looked like cartoons, but were based on the drawings inside, Anton explained. It seemed impossible to imagine this place, this land of nomads, as its first inhabitants had seen it. At ten thirty a security vehicle arrived, bringing the bad news we were secretly expecting. The workers were on strike, so the cave would remain closed for the next two days. It was the same at Font de Gaume, 'a solidarity thing', the guard said (or at least that's what I understood), but if we rushed, we might be lucky at l'Abri de Cap Blanc, where the staff had decided to stay open until lunch. We made it, just in time, for the last tour.

Technically, l'Abri de Cap Blanc isn't a cave, but a rock shelter, which the Cro-Magnon had repeatedly used during the last phase of the Upper Paleolithic – the so-called Magdalenian period, 17,000 to 23,000 years after humans had gathered in Chauvet. At the shelter's heart (and the reason we were there), a huge frieze is carved into the vault of an overhanging rock. Equipped with helmets, our small group followed the guide onto the now closed-off ledge. At first, I couldn't see much, except for the invasively bright exit signs and the hazy grey surrounding them. But then, there they were. A hollow space populated by creatures, many larger than us, silently waiting to be seen. The guide switched on a dim light and began to explain. The frieze appeared to be divided into two panels: ibex, bears, and some unidentifiable forms on one; bison on the other. I had to look hard to see these animals, but the horses, for whom l'Abri is known, were impossible to miss. Claiming their space throughout (at times overlapping and sharing contours), the largest, most daring, stood right in the centre: the only one facing east, its ears prickingly alert as if to defy or protect the others. Locked into the same stone, they felt like a clan, inseparable, yet strangely unattached. I couldn't tell if they were coming or going; how they related to the other animals above, behind, and beyond them. 'Change position and keep looking,' Anton whispered in my ear. 'Move slowly, one side to the other, and back again.'

From where I was standing, the herd seemed to be in motion, maybe running, but away from what? There – along the upper edge – a pack of (maybe) wolves was approaching, their movement directed by the stone itself. There was no terror. Just suspense. Acute, like the moment before lightning. As I crossed the space, this acuteness shifted into something less piercing, more watchful, eventually resisting. Seen from the other side, the horses were backed by a group of bison, bulging, but outflanked. How well must you know an animal to convince a stone to make it appear, move? Born into an animal world, the Cro-Magnon lived and died among these creatures. Hunters and hunted, continually shifting roles. How many generations of hands had carved these bodies? I wished I could touch them.

By the time we returned to the farmhouse, Martine was already up, baking bread, while Cécile, the calico cat, kept her company. 'We figured you two would be hungry. Sit down, coffee is almost ready. I want to hear all of it.' I fell asleep right there at the table, Cécile purring on my lap.

The next day, Anton's work in Chauvet began. I spent most of my time with Martine and Jean-Michel, helping them around the house and in the garden. After lunch they would take me on small hikes, telling me about their lives and the landscape in which we were immersed. How it must have looked 30,000 years ago, when there were only birches and pines and the Ardèche river ran up to the cliffs. 'We would be dead by now,' Jean-Michel interjected, referring to the cave lions and bears, mammoths, wolves, and rhinos that would have crossed our path to hunt and drink. The average life expectancy of the Cro-Magnons was twenty-three. The cliffs and the Pont d'Arc were already there, crossed by humans and animals alike. They also shared the caves, which might have been the primary reason why humans didn't live within them. Still, they delved inside to partake in unknown rites and rituals. 'You need to see a real cave, before Anton takes you to the replica,' Martine said as we were heading back. Jean-Michel suggested Gargas, which was only a two-hour drive away and, around this time of year, less busy.

While there are plenty of caves in the Pyrenees region, to date Gargas is the only one known to contain Paleolithic art. Older than l'Abri, but younger than Chauvet, it consists of two separate caves that originally weren't connected. The humans who entered one cave might not even have known about the other's existence, our guide explained. Flashlights in hand, we entered through a metal staircase, as if descending a pharynx, hollowed by water. The walls were scratched by cave animals, likely bears. Then, a huddle of small black dots. A single red one, further away. Like signs on a map. To encounter a cave painting always astonishes, no matter how often you've seen them in books or on screens. The silence is instant. Further down and deeper inside this body of stone, the profile of an ibex appeared on a wall above us. The red and black were barely visible, the creature as surprised by us as we by it. The walls curved and serendipitously revealed the black outlines of reindeer and bison, hiding in vaults. As we entered the second cave, which is wider but lower, I could see hollow spaces covered in red pigment, then charcoal. 'Take a close look,' the guide advised. There were halos (a greyish white, almost bone-coloured) within the pigment. And then I saw them. Hands. A cluster of hands, like ours. In the furthest chamber an entire wall was filled with them. The hands of men, women, children, pressed against the rock, while somebody blew or spit the pigment around them (aided by a straw-like tool, experts believe). Not all the hands lie flat. Some fold their fingers (perhaps they were missing), while others are bent as if shadow playing. The portrait of a clan. I waited for the rock to move and breathe.

That evening, a few members of Anton's team joined us for dinner. An eclectic mix of personalities, they included a geologist, a paleoethnographer, a geneticist, and an art historian. Over tagine and bottles of Cahors, the group discussed their research, which ultimately led to the quintessential questions that have accompanied cave art since its rediscovery in the early twentieth century: Why the Cro-Magnons painted, what it meant, and what purpose it served. The table was split into two groups: one group believed there was something shamanic about the paintings, comparing them with artefacts of other hunter-gatherer cultures; the more sceptical at the table argued, just as passionately, that for lack of evidence, we could never know for definite what they meant or what purpose they served. I quietly observed, and Anton (sitting across from me), every so often, watched me listen. When dinner was done and the team had left, he grabbed a bottle of schnapps and asked me about Gargas. 'So, what's your theory?' I said I wasn't sure, partly because I hadn't yet seen Chauvet. But the idea of shamanism (or some variant thereof) seemed to make sense. Especially considering that humans, to this day, perform rites and rituals to grapple with the natural and supernatural forces governing their lives. But something about the vocabulary – 'art', 'decorations', 'purpose' – felt wrong.

'You don't think it's art?' Anton prodded, his green eyes unwavering.

'What I mean is, by using those words, we refer to our understanding, not theirs.'

What did they call these marks they left in the dark? Did they name them at all? Whatever way they described them, I couldn't imagine it had anything to do with the desire to possess, adorn, or tame. To me, their images felt like traces of lasting presence. Perhaps they had a function similar to memory, to provide a kind of survival, not just of the painters, but also the painted. Perhaps our words couldn't get around them, humans or animals; the distance between us and them was too great. As we headed to bed, Anton told me that we'd be visiting Chauvet the following afternoon.

As a research member, he had arranged for us to see the replica without a guide, after official visiting hours, while the staff continued their work. I was surprised to find its interior less artificial than I feared. Even the smell and temperature (five to ten degrees cooler than outside) felt accurate. It reminded me of Gargas. Anton told me that everything had been reconstructed as exactly as possible, thanks to an army of graphic artists and painters. The cave was mesmerising, unlike anything I'd ever seen.

With its irregular, off-branching chambers, Chauvet sprouts like ginger root. Each chamber is its own cosmos. They vary in size and height, as they do in terrain, populated by fangs of calcified water. Recalling the colours of a body's insides (ranging from bone to tissue to flesh), some walls and floors are smooth, nearly damp-looking, while others are rough and bulging. Close to the original entrance, now buried by sediment, a stream of red. Dense and vibrant, trickling into red dots – almost squares. The marks were made by hands: their palms covered with pigment and pressed against the rock, as if the stone was touched from within. On the ground in the opposite corner, the petrified skull of an ibex. A male. 'You can tell by the form of his horns, the one part that never petrifies,' Anton told me. It felt as if I was being enveloped in a vast belly, shielding and frightening.

Ranging further, red hands appeared on other panels. They were joined by the outline of a rhino and the head of a feline merging with the body of a mammoth – fragmented yet fully present. The paintings gathered and dispersed like herds in a vast landscape. Between and among them, patterns of scratch marks, most of them made by bears' claws. 'This is where they went to hibernate,' Anton explained, as we reached the cave's largest and most central chamber, where many hollows cover the clay-cast floor. In and around them, the team found bones, including more than a hundred bear skulls; one of them sitting on a flat, plinth-like stone, deep in the cave. How would these early painters have known that the bears were gone when they ventured in? Spooked by this thought, I realised these horses, mammoths, and the small group of large insects (including a butterfly) must have felt like a benevolent presence – companions within the unknown, against the dark.

A wall of overhanging rocks. On them, the red fades and is replaced by charcoal and cream-coloured engravings, like a finger drawing on butter. Next to a huddle of swirls, an owl, long-eared with wings and feathers, turning its head, peeking down at us. In the Middle Ages, owls were considered harbingers of death. I prefer their symbolism in Greek mythology: couriers of wisdom and light. If an owl flew over Greek soldiers before battle, they took it as a sign of victory.

We made our way along the metal catwalk protecting the floor from our feet. The stone curves back and forth, and with it, the animals. A pride of lions streamed past, approaching bison who overlap and submerge within a crash of rhinos, tumbling into, then out of each other. No rules of perspective or scale seem to apply. A few exceptions aside, all these animals were drawn in profile, by different hands, sometimes centuries apart (even when sharing the same panel). What connects them is the sensuousness of their depiction, as if stone was just a different kind of skin. A cluster of five heads. Reindeer with delicate antlers, looking upwards, one with closed eyes, stretching its neck, as if to pick up a scent or enjoy the sunlight. Each figure bears a single likeness, an essence, like a portrait. The word that comes to mind is grace. A bison with six legs, hurrying forward. Then, horses. On one panel, two of them appear to walk towards each other, strong flanks and boisterous chests, never meeting. On another, a cascade of four heads with magnificent nostrils and bushy manes. All of them intrepid, their muzzles aligned in a flawless diagonal. The charcoal is thick and black, then rubbed into smudges of grey. Of all the animals summoned on these walls, horses are the only ones I've known up-close, and touched. The fear they can sense, the gentleness to which they respond. Domesticated, yes, but within them a kernel of wildness, forever untamable. Right here, I watched Anton crouch and draw.

'They don't know I'm doing this,' he whispered. Time is always too short in the cave.

'I think they do.' The intimacy of that moment. (How long did it last?)

As we returned towards the realm of early autumn, we passed smudges of black that seemed neither painted nor drawn. 'These are torch marks,' Anton said, 'to keep the fire going.' Stepping out into the misty evening, I thought about the dark these people delved into; a dark I cannot fathom.

'Do you ever forget why you're doing this?' I asked. Anton stopped and looked at me. 'I'd like to show you something,' he said.

Descending to the car, we took a small detour and climbed a narrow path further into the forest until we reached an opening in the cliff. 'Stay right next to me,' Anton said, grabbing two flashlights from his backpack. My breathing gave me away. I've always been afraid of the dark. 'Don't worry,' he assured me, 'there's nothing here. No danger, but no paintings either.'

This cave was much smaller, both in height and width, but it was long and windy. You couldn't see around the corners. 'I come here when I get out of touch, when research isn't enough,' he said softly, 'but our bodies always remember.' We walked a little further until the entrance was out of sight. 'Ready?'

he asked. Then, we switched off the flashlights. The darkness was so complete it swallowed us, defying words. My eyes tried to adjust; they wanted to hold onto something, but couldn't. There was no sound, no echo. I had never felt this blind before. Within this threshold of amazement and fear, I reached for Anton's hand, and I found it, reaching for mine.

The next morning, after breakfast, I said goodbye. Anton drove me to the bus stop, from where I headed north to Vézelay. His hug was warm and long. I reached the village by early evening and headed to the church, Sainte-Marie-Madeleine. Walking across the naves, I observed the drama of sinners and saints carved into the capitals, but I was really looking for the donkeys and oxen curving around the corners; one of them peeking from behind a giant leaf, almost hiding. The smell of stone and cold. I sat on a pew and watched dusk cast its shadows. Reaching for my scarf, at the bottom of my backpack, I found, nestled into it, a sheet of unruly paper, carefully rolled-up. Inside and written on a smaller piece (a scrap really), was a note: 'Tiny details imperceptible to us define everything.' – W. G. Sebald.

Dabs of charcoal. The heads of four horses. Retracing their lines, I thought of Anton's hand, and of the fingers who drew them first.

The Roaming Edges

Suzanne Walsh

I am looking at a moment, frozen. The trees on each side look like plumes of smoke. A wide icy avenue rises in a gentle ascent, small rounded trees like sentries. A dog crosses the snow from left to right, its body flattened to an ink-black silhouette, ragged snout surging forward.

In a culture increasingly dominated by imagery, eyes flicking hungrily across screens, I try and most often fail to remember the tracks I make. Each image I view online is loosened from its origins, set adrift to flow down endless feeds. I'd like to say I saw Josef Koudelka's photograph *Parc de Sceaux, Hauts-de-Seine, France* (1987) on a pristine gallery wall, or that I had some otherwise momentous encounter with it;

but the truth is I came across it a few months ago, while lying on my childhood bed, doing nothing much at all. I was hiding out in the valley I grew up in, needing rest from a year of intense work and family troubles. Instead of research for an essay I should have been writing, I followed trail after trail of images until my eyes landed on this photograph of an animal that has, in a moment, become something *else*. Outside, the last leaves of the year were sinking down into the soil and the birds were making their evening intonations. Whenever I called it up, the loping form of the dog returned to me, like an after-image. Like a black hole I could escape through.

The blackness of the dog in the photograph is as sharp as a paper cutout, loop of tail curled round like the faintly determined edges of the ornamental pond ahead, subsumed into the background. The peaks of its ears are echoed in the small trees. An agent sent ahead, on the scent of something unseen. I am reminded, not unreasonably, of Pieter Bruegel the Elder's painting *Hunters in the Snow (Winter)* (1565), in which returning hunters appear in the lower left-hand corner, walking towards a great vista of snow-covered fields, houses and ice skaters. Behind the hunters are the lean shapes of dogs, large and small, lithe bodies dark against the snow. Their tails are hung low, perhaps because little was caught. High up in the branches of trees that spread like cracks against a grey-green sky, black birds are waiting; a long-tailed one swoops across the sky, crossing the painting from left to right.

There is a figure in the distance, to the left, in *Parc de Sceaux, Hauts-de-Seine*, and I don't know if it is waiting for the dog or for us to see it. The drain underneath the dog, the lines of which add an extra horizontal note into the composition, suggests a force coming from below, a portal that has activated some kind of transformation when crossed. If you squint it also looks like the dog's blurred shadow.

It's late now and I'm back, again, sitting in my parents' house; they are both currently absent due to illness, leaving me here with the cat and the dog, the river estuary outside filling and draining. The house is very quiet; the land, not quite so.

I don't know what time of the year it is for you, reading this, but here there's still an occasional snarled bite to the wind, no matter what the ground and its enthusiastic awakenings tell me. Waking is a fissure that forms on the surface of sleep each night and breaks each day anew, some more easily than others.

Another dog that comes to mind is the sleek black one that attaches itself to the eponymous character in Andrei Tarkovsky's film *Stalker* (1979). This dog is one of the few creatures met by the main characters who are led into the Zone, a lush, mysterious, and unpredictable region with a dangerous sentience of its own. The dog follows like a shadow at first, appearing in dream-like sequences, but eventually returns with the stalker to the everyday world he inhabits, right into his household to dwell. He remarks only that he felt unable to leave the dog behind.

A friend reminds me, during a late-night Skype, that a copy of Bruegel's *Hunters in the Snow (Winter)* hangs on the wall of the library of the space station in an earlier Tarkovsky film, *Solaris* (1972); we see Hari, a replica of main character Kelvin's late wife, become mesmerised by it in one scene. To a disjointed, distressing soundtrack, her eyes track across the snow: the dogs, the skaters, the mountains, the birds, all in detail, all as though utterly alien to her own alien consciousness. The only version of the film I could find online was a version of the original Russian subtitled in Portuguese. I further translated this into an increasingly abstract English, until I had to abandon language altogether and just drift along with the imagery.

My father told me once about a black hound that used to knock on a neighbour's door every evening at six o' clock, long before I was born. Eventually the priest was called to bless the house and adjoining fields; but at the same time the following evening, the dog was heard howling in frustration outside the boundaries of the land. My father's friend told me about another incident that happened nearby, where a doctor was rushing to get to the side of a dying girl. As he was passing a small bridge en route a black dog stepped out and howled, blocking his way. He was too late to save the girl; later, he found out that she had died at the exact time he had encountered the dog. I asked him if the local greyhound track could have accounted for any of these mystery hounds and he give me an inscrutable look before throwing his head back in laughter.

I printed out *Parc de Sceaux, Hauts-de-Seine* so that I could look at it on the wall when writing; but the printer must have been low on ink or didn't read the file well so it came out like the blurred lines on a TV station improbably tuned, back in the days when a TV set was tuned, a blizzard of interference: something couldn't quite get through. I used my black pen to ink in the dog, returning it again to reassuring blackness, though the imperfect squiggles are obvious when examined closely. I felt the urgent need to return it to the power of its potency, its blank totemic presence.

At night, alone in my family home, I take our dog out for a last piss and see her disappear into the darkness at the edges of the garden. The water is lapping on the shore and the land is heavy under the stars, bright intense points holding it there, but only just.

The wind is blowing down the chimney and tearing strips of waves from the water's surface. I think about that exact place underwater where the river current meets the sea tide pushing up from the bay, powerful and driven, and how it pushes till sometimes it spills right out onto our lane. I wonder about being left entirely alone, someday, in this house.

Koudelka fled his native Czechoslovakia after photographing the Soviet invasion in 1968; since then, he has spent most of his years in transit. Famous for his images of that tumultuous time, his later works seem less concerned with humans than with the land itself. Looking at my notebook now I find fragments of things I've written down that Koudelka has said in interviews, about his life and work, and I see I've separated them from their origins, just like the images I view online:

I am still trying to find how far I can go ...
I realise the limitations ...
But in the same way the world is forming me ...
I might be dead myself by now ...
Nothing is permanent ...

Pure Animal Instinct

Vona Groarke

Meet Hoté, standing in my back field, lackadaisical. You can go right up to him, he won't mind: he's just that sort of beast. You can stroke his nose, offer him carrots, call him Eeyore insultingly, tease him about that small bald patch on his right rump – it's all much of a muchness to him. Look, you can even climb up on his back and imagine you're at the seaside again, and it's 1953, in Margate. Or Llandudno, since you never did see anyone lead a donkey with a child on its back along Spiddal Beach. (Probably all the Connemara donkeys were needed for hauling panniers of turf or leading a cart loaded with seaweed, or for posing for postcard photographs, if John Hinde was thereabouts.) But our Hoté has never done a day's work in his life. In fact, his existence is purely symbolic. I have no donkey. Hoté exists in name only, for the sake of a literary pun.

Imaginary animals are the very best: no vet bills, no shit to scoop, no walks in downpours in February with ruthlessly ebullient pets. I have several. Hoté, we've already met; Scat (a particularly querulous marmalade tabby); an Alsatian called Lorraine; a lamb called Shush; and Greasy Spoon, the Friesian calf. I used to have more but there's only so many imaginary animals you want about the place. After a while, they can't help themselves, they bicker, wishing they were real so they could skedaddle, so they could jump the fence of me.

Mine are would-be paper animals. Not exactly origami since that suggests physical entities (3D, angles, folds), whereas mine are inherently abstract. It doesn't matter if they have a leg to stand on, they are creatures of language, the closest to puns I ever hope to come; punkeys, if you must. Before this, my notional animals were just that: here, they are committed to print for the first time, tethered by a thin black line to the stem and stroke, the joint and crotch, the legs and shoulders of whatever font you're reading now.

Thus is the abstract brought to book; the slippery fantastical noosed through the nose and slip-knotted to a surface that thinks it is a page. If it were a page, I could maybe have drawn here a cheeky-but-recognisable representation of my private creatures, as those long-lost scribes did in the Book of Kells, with their pretty wolves and paste-like peacocks, their sculpted horses and serpents like feather boas round flappers' necks; their effete cats and agitational dragons; their chirpy pigs and tail-biting dogs, even their hard-working calves (depicting, reverentially, St Luke, but nodding also, I suspect, to a throughline of form and content, a calf drawn on calf-skin vellum being an off kind of slant rhyme). Any of these punkeys I could draw for you, if I had the *Children's Colouring Book* version of my imaginary bestiary.

But, of course, I could not draw them for you. My punkeys are entirely lacking in physical specificity. Is Hoté brown or grey; male or female; bony like a rotting ship on a tourist trail, or fluffed up and hearty, bellyful? Who cares? And what matter my (lack of) drawing skills? My punkeys don't require beaded tails or nicely blue-inked torsos. I need buy no lapis lazuli pigment nor brushes made from the hair of a kolinsky: particularity is useless to them. Donkey Hoté is a one-trick pony: I don't need to flesh him out, because I don't need you to believe in him; you've already got the joke, and the joke of his name (if it's even that) is the height and width and end-stopped point of him.

This naming of animals is a tricky business. I grew up on a farm where a rather Wordsworthian character tended the livestock. He had two nephews, one donkey, and a somewhat disputatious personality. If he was fighting with one nephew (let's say, PJ), he'd call the donkey, loudly and in no way gently, 'PJ', until it was the other lad's turn. He seemed to fall out with them on a rota and to take his annoyance out on a beast that was a sorry excuse for animosity, whichever way you looked at it.

When my father decided we should get a new dog, there was ructions in our house about the name. The two youngest (including me) were nominated as namers-in-chief, but some of the older, more worldly, siblings decided to try to influence the choice. We were privately and separately co-opted to call the dog Randy, because that would be funny, hollering 'Randy, Randy, Randy' from our house into the environs, day and night. The campaign was discovered, Randy ousted and, well, meet our golden Labrador, Monty, named for no reason I can remember except maybe the very worst reason of all: it was unexceptional.

My late dog, Judy, was named for another childhood dog we had, a docile but somewhat regressive bitch who flat-out refused to learn tricks. (That dog I could draw for you, perfectly, because she was a dead ringer for the newer Judy, the one my son and daughter picked out of the Winston-Salem Animal Rescue, four decades on, who travelled across an ocean with us and who was the sweetest presence in our lives, despite her only tricks being eating and breathing which, in fairness, she accomplished effortlessly and elegantly.) Nicky was the trickster: he'd jump through a Hula-Hoop, no problem, and get up on his hind legs like he was Rudolf Nureyev (or Rudolf Nureyev's dog). Another trick he had, devised all by himself, was a patient Sunday-morning cower behind the hedge until any old lady (and it was always old ladies) would happen to cycle by to Mass. Out he'd spring, barking a fury and nipping at legs, and if he got her down off the bike, we – watching from an upstairs window – would take the kind of pleasure any amateur misanthrope would take from such a display of expert malevolence.

Nicky's personality must have been nominatively determined by his devilish name, though how we could have foreseen in that mewling, part-collie pup a dog of such splendidly callous intelligence and anarchic bent is beyond me. But of course, Judy was always going to be sweet, just as Nicky was going to be the opposite, and Hoté too knowing and notional to saddle with actual donkey flesh.

Enter, the pantomime horse of metaphor: front end, poetry; back end, wellspring, fund of reality. Like any good horse or '20s Hollywood actor, the pantomime horse keeps ever one foot on the ground. Unless it is bred from Surrealism, it will fix itself with a tether of words to the Real World, and its stuff. Those curious-drawn creatures in the Book of Kells are beguiling insofar as they are recognisable: strip the wolf from its intimations of stalkerish wolfdom, and the drawing becomes a doodle, whimsical or winsome, sure, but probably not much more. There is a proposal and there is a fact: the fact is 'wolf'; the proposal is 'furbelow', ink and slink translated together into suggestively metaphorical design. In connecting my dog Nicky with Lucifer, I metaphorise him: I ascribe to him, in words, a personality that might be historically accurate, or might not – it hardly matters, really; Nicky is a creature of the past. But I've made him a recognisable Bad Dog for the sake of this essay and my strategic intentions, my interest in how animals can be herded into creatures of language. If he were alive, I'd forgive him for taking a chastising nip at my own ankles, or at my metaphorical feeding hand.

In taking to poetry, I converted almost every aspect of my life into a papery version of itself. That's what we do, we poets: we feed experience, headfirst, into the poetry wood-chipper. In goes the present, past, and future; out comes the flittery bits of language we decide we will call poems. We seize our chances as they fall, lighting on whatever materials to hand to offer a credible version of the selves we project, like shadow-puppets, on the white space of the poem. And poems love animals, light on them as if they were poor beasts saddled with a doghoused nephew's name, to be shouted at and prodded with a stick into subject matter. Something about their fund of innocence and insouciance, I'll hazard, as well as their potential for being flipped, nice and easy, to metaphor.

If I've a mind to, I can pick up any anthology and find a litter (or a kennel or a nest) of animal poems. It's a curious magic trick, this converting of flesh and blood animals into language ones. (It also works for houses, lovers, plans, and cash – try it, it's really simple: you just have to keep practising.) Some win me over more than others. Eliot's cat poems I find scratchy and fluffy, but Ted Hughes's 'The Thought

Fox' slinks into a culvert in my thinking about writing now and then. Likewise Heaney's 'The Skunk' (or 'The Otter', or 'The Badgers' – all nustling up together in *Field Work*), where the animals deftly and delicately rummage in imagery to discover and dislodge a human truth. Whereas Lawrence's 'Lizard', with its smug aphorism, I'd happily drown in a critical barrel. Bishop's 'The Moose' lurks, magisterially, over my road into writing poems; and Yeats's hare's collarbone (surely the strangest collarbone in Irish poetry?) still has me pressing my eye to its gimlet hole, wondering what I'm seeing.

My favourite animal poem-moment is from Frank O'Hara's 'Animals', from 1950, a poem about company and, best of all, the company of young people when you're young; the greedy ease of it.

> Have you forgotten what we were like then
> when we were still first rate
> and the day came fat with an apple in its mouth

... goes the first stanza, a question that is never end-marked, for to finish it so would be to suggest that this question's over now. Asked and answered; complete and completed. Whereas, of course, we know that the answer, were we honest enough to give it, would have to be, 'Yes' and, at the same time, 'No'. Which is how we remember our young selves, in partial snatches of punishment, indulgence, judgement, brutality, and sentiment. Which is pretty much how we relate to animals, too, even ones we love.

(Maybe that's why we're fond of animals, because they remind us of our youth. Or rather because, even more so than our own children, in whom we – mostly – tolerate a degree of independence, we behave to animals pretty much the way the world behaves towards youth; as regulators, owners, herders, consumers, buyers, and sellers; as *Boss*. Rewards for obedience, punishments for cheek; no wonder any given generation despises the one (or two) before it: we deserve it, certainly (though not always for the same reasons youth thinks we do).

Our own youth, dumb insensate, had only the loosest idea of how beautiful it was and how short-lived. Its needs were animal needs and its intelligence largely unconscious, instinctive, which makes it vulnerable to abuse. And by god, do we abuse it. Cherished and loathed; blamed and iconised; we thrash it to within an inch of its life and then we bind it in gilded chains, and pat it on the head.)

These days I try to resist writing poems about animals in the same way I try to resist poems about my being young, and for the same reason: neither can argue back. It wasn't always so, but I've already written my dog poem ('This Being Still' in *Double Negative*, since you're good enough to ask), and I will never, hold me to it, write a poem about a mere cat. While poems are excellent tools for many vehicles (love, loss, rage, desire – the full gamut of feeling, crossed with conviction, scored with style), I think I might just choose to take my animals metaphor-free. Which (if the opposite of metaphoric conversion is unmediated actuality), is to say, real. Shitting, slurping, snoring, demanding, devoted, cross: my future animals (as Archibald MacLeish decreed about a poem) – 'should not mean, / But be.'

But they won't be. I could as easily say I try to resist writing poems about animals in the same way I try to resist owning one. I live a solitary life. Sometimes I imagine, of course I do, that solitude leavened by another breathing creature in the room. A dog at my feet; a pair of eyes seeking out my own; a shape to the day, with walks and meals and a reason to get up at such a time; a reason to speak, to exercise a voice in a way that doesn't make me sound batty to neighbours; a way to receive and return affection and, thus, to maybe get back again to halfway liking the world. There's a spot on the floor in front of the stove that would be ideal for a dog, I already know it, but if she chose to sleep at my feet, that would be fine too.

Mostly I'm fine with my chosen solitude. Mostly. But it's exactly the sliver of me that isn't (that sometimes rises, like a good yeast loaf), that rules out such a plan. I feel the occasional need to flit; to spend a month or two in some other place, eating food I don't usually eat, talking different talk (any talk). And even if I never go, I find I need the possibility of the going. I don't want to be tied to one place only, like a tethered donkey, on howsoever long a rope.

Hence, Hoté in the back field needs to be the kind of donkey that doesn't need any minding, that can take care of himself. Just as Beryl the short-legged

Basset Hound will need to forego walking, eating, or the need for prohibitively expensive kennelling. It's a stark choice: procure an animal and stay home; or be free to go wherever, whenever, and forfeit animals.

I could say that Hoté – the set of him, the need for him – transcends the role of mere punkey to pretty much infill the colour and shape and depth of loneliness. Which doesn't make me like him less; quite the contrary. Hoté underwrites the life I choose, with its various heres and theres. He permits me, insouciantly, to avail of the insouciance I confer upon him. Which turns out, of course it does, to make us both more real to me.

Say God, Say Bird[1]

Latifa Akay

If you ask me who my God is, on whose name I call. For me, the song of the blackbird is intertwined with this line sung by Yusuf Islam, in his post–Cat Stevens era. At home growing up we had a cassette tape of Yusuf Islam reading prayers and singing Nasheeds, including this one. Occasionally moments of bird chorus feature, the confident song of the blackbird coaxing the other voices in, one by one.

Pure, full-bodied, reverberating. I feel myself sitting up straighter. Something like hometown pride. I have heard so many blackbirds in my life, and yet when I pause to think of its call, this song

1 After *Say Love Say God* by Zaina Hashem Beck

often comes to mind. It's funny isn't it, associations. Say God, say bird. If I were to make a mixtape of my childhood, the song of the blackbird would be on it. I imagine the sound cutting through the heavy brown-gold curtains of my childhood bedroom. Piercing the single-pane windows. There I am, wiping the condensation off the glass, peering out. Appearing and disappearing in my own breath.

The windows of my childhood were always ready to weep. You touched the cold glass, and a droplet would immediately race to the wooden frame. I want to stand, arms around my younger self, and watch a blackbird covering the lawn at pace, hop hop hop – it's away. This bird is fast, it's like the Golf GTI of British birds. Or as Seamus Heaney put it:

> On the grass when I arrive,
> In the ivy when I leave.

Is it stereotypical of me – a person from Belfast – to love the song of the blackbird? Probably, but that's okay. It's okay to be sentimental at times, to say you know what, I love that place. The smell of the soil, the bite of the mornings. In London the birds are quieter. I rarely hear a blackbird sing in Tottenham, but this is the place I choose to live. Often people choose to live in one way and long in others.

In Tottenham the sound of parakeets cuts through the most. They move in fast-paced screeching flocks, like they are passing on something important. They are the gossip girls of the London bird scene. Or maybe that's the starlings, or the sparrows, or the crows. At nighttime the parakeets settle in their thousands in the big trees along the River Lea. Chattering until darkness descends. There are several parakeet origin stories in England now. Some say Jimi Hendrix released two of them on Carnaby Street, some say they escaped from the set of *The African Queen* in the '50s. Others say they are most likely pets that got away or were released or abandoned. What will the British people do if parakeets become more prevalent than pigeons? Will it sit uncomfortably with them if Trafalgar Square is full of green birds instead of grey ones? Will they replace the word 'flock' with 'swarm', 'song' with 'racket', 'group' with 'gang'. Will they say *hang on a minute; the pigeons are our national bird. They carried messages for us in the war. Now they're being pushed out by parrots – SPEAK PIGEON SPEAK PIGEON.* Truly all we can do is at least expect the goal posts to change.

I want to take you back now. I see myself peering through the binoculars I got for my tenth birthday. I remember the card: square and shiny with 10 in big multi-coloured double letters. There is a disagreement at the dinner table that night and the kitchen is full of shouting. We are eating fried eggs, chips, and beans, my birthday meal of choice. I cut my fried egg into tiny pieces.

Later, I sit out of the way and thumb the edges of the 10 on my card. Eggshells everywhere for days.

I'm walking on eggshells, wooaaah
I'm walking on eggshells, woaaaaah
I'm walking on eggshells, woaaaaaah
And don't it feel –!

10, 10, 10, 10, 10. Tiptoeing in my mind.

What do I see through the binoculars? Collared dove, great tit, blue tit, dunnock, wren, wood pigeon, green finch, chaffinch, often blackbird, mistle thrush, song thrush, often magpie, chaffinch, rarely bull finch, rarely jay, often robin, coal tit, starling, magpie – you can add more. The Antrim Road bird population in the '90s. Searching for a black cap, searching for a wax wing, searching for an owl, any owl. I see an owl in a dream sometime around this age. It flies low over the back garden, big in the daylight. I wake up in the top bunk disappointed. This is the only time I've ever seen an owl in flight in the wild.

Popstars are not allowed on my bedroom walls, so I have posters of birds. The long throat of a black swan, the pert tail of a wren, the fanned feathers of a peacock, the sleek sides of a mandarin duck. Birds > boys. As an adult, friends tease me about my birdwatching past. Birdwatching, eh? Birds with feathers, I protest. I think of Gerard Manley Hopkins's words in the poem 'The Windhover', dedicated 'To Christ our Lord':

> My heart in hiding
> Stirred for a bird, – the achieve of, the mastery
> of the thing!

A few years ago, we got Baba binoculars for his birthday. He keeps them on a small shelf within reach of his armchair in the living room. When I visit, I always take the seat on the sofa opposite. I watch him reach for his binoculars. He complains about how fast the birds move. I tell him that when I hear sparrows cheeping on a hot day, I am transported to Ankara. Close your eyes, I say, you see what I mean? He sends me photos and videos of the sky from his armchair when I'm away. Sometimes he asks me, why is the female blackbird brown? Does she have an orange beak too? Why is the beak orange? Why call it blackbird if the female is brown? I say, I don't know, I just don't know.

During the pandemic lockdowns, I start to enjoy birdwatching again. In autumn 2020, I am offered a pair of binoculars by my friend Deborah. She tells me that an elderly friend from church, Jocelyn, has asked her to find a good home for them. We meet in Hackney Marshes on a drizzly morning a few weeks later. Deborah is giddy and energetic, dewy droplets making a halo out of her grey hair.

Jocelyn's with the angels now, she tells me, searching in her bag for the binoculars. I stare at her, hang on? Yes, she passed away yesterday, she tells me straightening up. Yesterday, she repeats. But she was at peace, she was ready. While writing this, I realise that I have forgotten some of the things

Deborah told me about Jocelyn. I remember that Jocelyn was from Antigua and loved dancing. That she lived high up in an apartment block in West London. 'A little light box on top of the city, she would watch everything from up there,' Deborah said. I message her and ask her some questions. She replies:

> Here is a pic of lovely Jocelyn in her best Antigua T-shirt and oft-worn red hat, dancing at some event or other. She did indeed live in a light box, not exactly a tower block but a few floors up. One whole wall of her studio was window, and she often had her bed right in front of it with no curtains. There was a time when she had another wall painted as an Antigua beach scene.

I study the photo of Jocelyn dancing, her eyes are closed, her skin stretches over her cheekbones. Her T-shirt is lime green, in the middle it says 'Antigua, Beautiful Paradise'. Around the words there are embroidered flowers and hummingbirds.

Jocelyn's binoculars don't focus well, but they still offer something of a different perspective. If you pass them on, tell whoever you give them to about Jocelyn, Deborah says. At some point soon after I buy a pair of new binoculars because I've decided I'm taking this seriously again now, but I keep Jocelyn's stored safely. I will give them to my friend Asma who I know will carry on Jocelyn's story. I realise in writing this that I still need to do this.

I take my new binoculars to the Walthamstow Reservoirs one day in winter and find myself caught up with a group of birdwatchers. Before I can remove myself, a wiry white woman in wellies tells me to keep up, then points to the sky. *Up there, folks, peregrine falcon on top of the pylon!* I raise my binoculars dutifully and see it sitting there, Hopkins's 'dapple-dawn-drawn Falcon' in still life. Broad shoulders hunched, waiting. It's the first time I have seen a falcon in the city.

Around this time, I cut out a picture of a lithograph by Lubaina Himid from a magazine. In the middle of the image a kingfisher and a robin sit on either side of an azure blue triangle with a flower print on it. The words 'Birdsong Held Us Together' are written in a greenish border around the four sides of the image. Squares of the same deep blue with prints of birds on them mark the corners of the image like stamps. I scan a copy and frame it for my mum. She has it in her bedroom. This is the room that she prays in. In the summer she prays Fajr and Maghrib with the windows open, the dawn and evening bird chorus surrounding her in prayer. I hold her there in my mind, my mother praying in the breeze and bird chorus, lips moving in *sujūd*. Say God, say bird.

During the pandemic I end up staying with my parents in Belfast for three months. I plan to initially stay for three weeks, but I extend my stay two weeks at a time until three months has passed. We walk along a road called Sandy Lane every weekend and it is there that we discover the buzzards. Two of them, either sitting on the roof of a small, abandoned outbuilding or in the branches of a tree nearby. It becomes a project for the three of us: *No sign of the buzzards. Oh hang on wait, there! Where? There!*

On one of these walks, Baba says, I'm glad I didn't wear my top with no arms, what's it called, giblet? *Gilet*, I say. How do you spell that? he asks. G-i-l-e-t. You don't pronounce consonants at the end of the word in French, my mum says. You wonder what's the point in having them there, she adds after a pause. I think of a photograph of me on my eleventh birthday, dressed in a khaki-coloured gilet and matching trousers. I stand in the hall, hands behind my back, chest puffed out proudly.

In Farid ud-Din Attar's *Manṭiq al-ṭayr* (*Conference of the Birds*) a large group of birds, under the leadership of the hudhud (hoopoe), undertakes a voyage to find their mystical leader, the legendary Simurgh. After a long and perilous journey, only thirty birds remain. When they finally reach their destination, it is revealed to the birds that what they were searching for was within them all along – Simurgh means *sī* (thirty) and *murgh* (bird). Attar's birds are disciples, the hudhud a spiritual guide, and their journey an allegory for the 'Sufi way' to awakening. This is of course the same hudhud who spoke to Prophet Sulaiman and the Queen of Sheba. You'll understand then that I felt something akin to starstruck when I saw a hudhud for the first time. A bird of unextraordinary size with a regal headpiece wandering across the grass in Jinan's mum's garden in Sharjah. Then – during a trip of only a few days – I saw hudhuds in many other places: stretches of lawn, grassy verges, mall car parks. I understood on this visit that a hudhud in Sharjah is perhaps as commonplace as a blackbird in Belfast.

I have a pdf of the translation of *Manṭiq al-ṭayr* on my computer, but I have only read fragments of it. I believe this is a book that should speak to me, but I haven't been moved by it in the ways I want to be. Do I trust the translators, do I trust myself? I raise this with my friend Ghazal when I next see her, ask her what she takes away from it. It has been a while since she has read the book but after brief consideration, she says that there was something about the 'self-limiting beliefs' of the birds that stuck with her in the story; their fears of whether they are good enough to meet their maker, their reluctance to step out of their comfort zones to embark on the journey. The heron who says: 'The Simorgh's glory could not comfort me; My love is fixed entirely on the sea,' the finch 'a nervous flame' who believes itself to be too weak to make the journey, the owl who would prefer to remain in abandoned ruins that 'harbour buried treasure': 'Love for the Simorgh is a childish story; My love is solely for gold's buried glory.' When we think about it, Ghazal says, we allow a lot of obstacles to get in our way. God says to us in the Quran that they are closer to us than our jugular vein, but how much do we allow ourselves to believe that?

Weeks later we message about the text again and Ghazal says: 'After that conversation I embarked on listening to a series of lectures and translations of the poem in Urdu and it feels so much more satisfying. It quenches like water after the syrupy English translation lol.'

Post-lockdowns, I find that I'm unlikely to choose birdwatching when there are other activities on offer. But I am tuned into birds in a way that feels old and new at the same time. Jinan sends me videos of starlings talking, which blows my mind. I showed her how to identify a starling on a walk we took during one of the lockdowns and then she discovered through TikTok videos that starlings can mimic people. According to some, starlings can talk better than parrots.

My earliest memories of starlings are in my primary school playground in Mossley. In my memories there are hundreds of them. They are always on the concrete ground, glinting under the greyish yellow Belfast light, comfortable among the fall and thump of children's feet. Later, for years on our journeys home from secondary school on the M3, we would pass starling murmurations shapeshifting over Belfast Lough. I don't think I've seen a starling murmuration since that time.

There, there! my mum would say, pointing in the direction of the Odyssey as we passed over the water. Giant clouds of black dots dipping, soaring, swelling. Suddenly turning inside out like a sleeve in a jacket or an umbrella on a windy day.

Three Dogs and One Cat

Honor Moore

I come from two families who raised dogs, which was a kind of sport or vanity business for a certain class of wealthy Americans in the last century. My mother's father raised long-haired fox terriers, and then during the 1920s his kennels became the first in the United States to breed Afghan hounds. Picture this: luncheon guests gather to marvel at the spectacle when the gates are opened and gorgeous golden animals leap across acres of very green lawn.

Dogs lived inside the house or out, a canine version of upstairs downstairs – intimacy versus property. Intimacy: as a girl, my mother was given a 'dark brindle' Afghan (aka Badshah of Ainsdart) named Loppy as her pet, and her sister, my aunt, a miniature boxer apt to leap onto the dining-room table and eat the butter balls. Property involved the use of words like 'bitch' to describe a female dog and was connected to breeding, and a dog with extreme illness or behaviour problems was 'put down'. There was a tone of voice my breeder grandparents used with dogs, a kind of brutal almost-shout. I remember my grandfather's otherwise-restrained second wife, clad in riding jacket and jodhpurs, whippets gathered around her, scaring me when she loudly tamped their enthusiasm: Down! Down!

My other grandmother had indoor poodles all through her eighties, but when I was a child, she raised dalmatians allowed only outdoors. All their names started with D, which designated them descended from Dipper, the patriarch who had brown rather than black on white spots. There is a photograph of me in my grandmother's garden, a little girl, dreamy-faced in a plaid shirt and shorts, three dalmation puppies on her lap. At home, Chelsea, a long-haired dachshund, was on the scene before I was born, and later a Scottish terrier called Smitty, but at my grandmother's farm on weekends, I appropriated Digger, black and white and one of Dipper's sons. My father had already read us *The Lion, the Witch and the Wardrobe*, and I was convinced that if I focused intensely enough, Digger would talk to me and I would finally have a soulmate.

He did his best, following me when I rode my blue English bike, walking with me in the woods near my grandmother's big house, being present warmly when I stroked him or looked deep into his eyes. But he never did speak, and even when I tried to imagine what he might say, he remained silent. After we moved to Indiana, we got to the farm only twice a year, and I saw Digger get old. It wasn't that he didn't recognise me, but in the way of animals he was ageing fast, and we were no longer contemporaries.

In Indianapolis, I begged and begged for a dog for my twelfth birthday. Our neighbours had what I called 'a beige cocker spaniel' and I wanted one too, even though my parents, especially my father, didn't like the breed. An adorable puppy arrived, I named him Percy, but I was left – oldest of seven children with busy parents – to take care of him by myself. Unlike my parents who had grown up with dogs, I had no idea how to train him. He wouldn't 'stay', he wouldn't 'sit' and unlike Digger, he didn't seem to understand me at all; in time, he became my brother's dog not mine. I felt guilty and sad, but I was soon taken up with my own life. One day when I got home, I called and Percy was nowhere to be found. Where's Percy? I asked. I can't remember who of my family laughed. He had bitten the postman, they said, and had been 'put to sleep'. I burst into tears, but they didn't understand why, they said. Percy had been gone three days and I hadn't noticed. I have not until this moment wondered what was going on for Percy himself: Did he feel abandoned by me? Was my brother just more fun?

I did not again recognise a friend in an animal I 'owned' until in my twenties when a big orange cat named George turned up at the house I shared with my boyfriend at the seashore. George's family had acquired a parrot and he was so offended he'd adopted us, was what we said. He had a diffidence which made that story plausible, but perhaps our attraction for him was the tuna we put out for him every day. In any case, when September came, he accompanied us back to Manhattan.

George was not one of those cats who gains weight and becomes big and round; he was large, leonine, and lean. I had Victorian furniture upholstered ocean blue, and we arranged the armchairs in a circle. At one tea party I remember, George took his decorous place on a chair and sat calmly, as if he knew the gathering was important, which it was. The great Adrienne Rich was there with her partner the novelist Michelle Cliff, and the discussion was of poetry and feminism. It seemed to me that George was following the conversation. When the boyfriend and I split up we divided custody, and ten years later, when George contracted irreversible kidney disease, it was my week and I who took him to be put to sleep, holding him as life left his body.

Until I met Brussels more than a decade later, I had no more animal friends. (I have never liked the word 'pet'; it seems belittling.) For twenty years I lived in the country in northwestern Connecticut, and one day in the early 1990s, a friend called me. I was in New York overnight.

'Your dog is waiting for you,' she said. 'Stop on your way home at the shelter in New Milford.'

I wasn't sure I wanted a dog, but I followed her instructions.

The friend was the owner of Brecon, a spectacularly charming border collie named for the Brecon Beacons in Wales. I didn't yet know that the Brecon Beacons were a mountain range, but I was lonely and I liked the idea of having a beacon of my own. A dog would be 'good for me', friends said. As I think about that time, the word that comes to mind is yearning. But what was I yearning for? Recently I have identified an enclosed feeling I get when I'm overwhelmed, bored, tired, or angry. When I don't feel that enclosure, my life seems full of magic – people are magic, food is magic, writing goes well, and my sleep is full of magical dreams.

I stopped at the shelter, gave my name, and was led to a cage. The dog, a bit bigger than Brecon, was silky and all black except for a sprinkle of white on one front paw and a beautiful snowy neck scarf. I loved her immediately. She jumped around as I took her – the attendant said training her would be challenging, that she was 'a handful'.

'Do you have an idea of her breed?' I asked.

'A Belgian mix.'

Immediately, I named her Brussels. I soon learned that Belgian sheepdogs have a history. During the First World War they heroically carried messages across enemy lines. They were skillful and calm and their black colour made them invisible in the night. Once when I told an older friend that Brussels was a Belgian mix, she scoffed, 'Don't be ridiculous, she is too beautiful to be a mix.' I didn't care about purity, but I loved the idea that my dog was descended from war heroes, that her ancestors had affected the course of the Great War. I couldn't prove that she was fully a Belgian, but she was the right size, had the right coat, and the exact markings, the white salt sprinkle on one front paw, the white at her throat.

Brussels became a new source of magic. The idea of taking a walk might flash through my mind and she'd noisily race through the house right to where I was sitting, ready to go. I began to believe that we communicated in brain waves. I thought of her as related to crows, starlings, and black cats, as a piece of the night. I couldn't get over the softness of her ears and the depth of her dark brown eyes. I got used to her as one does to a longtime friend, meaning I got to *know* her. One of the ears bent slightly which gave her a quizzical air – when she looked at me she always seemed to be asking a question, the kind of question you can't really answer, like 'What is the meaning of the universe?' or 'Is there such a thing as eternity?' I can hear my dog-breeding grandparents: Don't be ridiculous, she's just asking for food.

I can still summon her quiet attentiveness. I became charmed by her intermittent disobedience and independence of spirit; her only consistent obedience was to my question 'Hand?' at which point she'd put her paw in my hand and I would squeeze it, and she would squeeze back. We often did this as I drove and she sat next to me, in the passenger seat. Our worst fight was early in our life together when she ran out onto the road during an ice storm. I ran after her, my heart pounding, already crying as I yanked at her collar. I knew she was resisting because she was scared, but something in me snapped and I began to whack her, shouting outrageous curses as cars slid by barely visible in the sleet. This was not her fault, I realised as I pulled her towards the house. I never forgot the look in her terrified eyes, and later, when I thought of that day, I'd rub her soft ears and say, 'I'm sorry, so sorry, but I didn't know another way to save you.'

Another time, on our morning walk, she ran away and I didn't find her for three days, I drew a portrait of her in black Sharpie and put posters up all over, and one morning the phone rang: 'She came to our door,' the man said. I grew to trust her wanderings. There was an afternoon walk we took across the road to the river, and finally after an hour looking and calling in the brambles, I decided that if she wanted to risk her life by crossing the road alone, it was up to her. She had a luscious wet nose always, and aside from stroking her ears, I'd stroke her neck until the two of us locked eyes and she'd gulp as if she were a bit self-conscious about how good it felt. I was shy too, I told her in brain waves, and not easy with intimacy.

When Brussels was fourteen, we moved to Manhattan. Walking her in my Upper West Side neighbourhood, I came to rely on her beauty, which was so great that people stopped us on the street to comment, and she would calmly absorb the praise. When she fell ill at sixteen, a dog-owning friend said, 'We are their keepers, not their keepers-alive.' Brussels was old for a big dog, I was told, but she still had a buoyant stride, still jumped towards me when I opened the door and still attracted attention on the street. And then she got really sick. She became incontinent but, ever considerate, relieved herself only in the kitchen to spare my rugs and wood floors. One day, sitting on the floor in my bedroom, she looked at me with an almost violent gaze: *Get me out of here*, came the brain wave.

I didn't want Brussels to die in the city, so I called and made an appointment and drove us up to her old vet in the country. When we were called into the examining room, I asked her one last time to lie down and she did and the vet administered the sedative which precedes the annihilating shot. As she was going under, she put her beautiful and familiar paw into my hand and I squeezed and she squeezed back, and then came the final shot and she was gone.

I now remember another command Brussels responded to – when I first got her I saw that she could spin and so I took to saying in my sing-song, only-to-dogs voice, 'Dancing dog!' and she would spin. The first time we did this, it was late at night and the snow on the ground was lit up by a full moon.

In a photograph I took, which I have framed on a bookshelf, Brussels sits in the doorway of the back porch, looking up into the forest that rises from the brook at the base of the lawn. There were squirrels in that forest and also deer. Once in a while Brussels would take off like a shot, leap the brook and dash up the hill in pursuit of some crackle in the woods only she could hear, a waft of scent only she could smell. But in the photograph, she is waiting. It's in black and white, and she looks like a silhouette. Her abundant tail rests on the brick floor to her left, merging with her shadow. It must be fall because the lawn isn't smooth and beyond her are bare trees, vertical on the upward slope, casting their own shadows. It is in this photograph that Brussels continues asking me her questions: 'What is the meaning of the universe?' 'Is there such a thing as eternity?'

Michael

Erica Van Horn

7 August

An elderly robin has become a friend. He stays nearby whenever we are outside. Mostly he sits on the back of the chair where one of us is sitting. Then he moves to sit with the other one of us. He hops along the tabletop. His head and wing feathers and his red breast are scruffy looking. That is how we know he is not young. His scruffiness is what makes him distinctive. It does not matter which table we are sitting at or whether we are drinking tea or coffee. He seems to like the companionship. Or maybe it is the sound of our voices.

MICHAEL

9 August

The robin has a name now. He is Michael. He joins us by the back door and he sits with us at the large table over by the fence. Wherever we are out of doors, Michael appears and is committed to Staying Near. He sat on a branch while I picked raspberries. He came into the kitchen and rested on the windowsill while we were preparing food and cleaning up. He jumps when there are sudden movements or loud noises but he seems to enjoy quiet words or nonsense syllables babbled softly in his direction. We are not getting too much done because we are constantly popping in and out of the house or looking out a window to see where Michael is. We crumbled up crackers on the table for him. There is water for him to drink. He came with me as I went all around the tree and up the stone steps squeezing figs to test them for ripeness. He appears to pay careful attention to every single thing that is done out of doors.

11 August

Michael was late arriving this morning. Every day he has been waiting on the table outside the kitchen before we are even up. He was doing the morning waiting for a long while before we realised that it was the same bird waiting out there every morning. Today there was no sign of him until noon. We were worried about his leg. We are still worried about his leg. One leg is now at a wonky angle. The displacing of the leg happened some time yesterday. A bigger fatter stronger robin had been rushing onto the table and chasing Michael away each time he was there. That was when we realised that Michael is not an old bird as we first thought. He is a very young bird. The older robin who was pushing him out of the way had seniority. At the end of the afternoon we saw that Michael's left leg was sticking out at a right angle. He kept falling over while trying to eat crumbs. We thought he flew at a window and knocked himself down onto the ground. But now we wonder if the older fat robin chased him and frightened him and forced him into the window or into the wall of the house. It must be the impact with the ground that damaged his leg. We were happy to see him back today but we are worried about the leg. We are worried and we have no idea what to do about it.

15 August

Michael is here to greet us. Michael greets us whenever we return to the house. It does not matter if we are coming back from ten minutes away or from three days away. He arrives and hovers close and comes indoors and generally lets us know that he is glad to be nearby. I have now been told that it is normal for young robins to adopt people. I thought we were special and that Michael was special. I still think he is special. It's just that it is not such an unusual thing to have him want to be with us. I do not mind that. Michael flies away when the other robins come to frighten him but nothing we do frightens him. He has graduated to sitting on Simon's shoulder now. He has not sat upon me yet but he is happy to sit very close to me.

18 August

Michael sat on a low leaf while I collected figs. His leg looks much better. He still favours it, but it no longer sticks out at that terrible angle. He sat on leaves or on large stones while I picked raspberries. He has no interest in eating fruit. Maybe robins do not eat fruit or maybe he is just too young to know that he might love it.

26 August

This morning Michael is sitting on the table while he eats his crumbs. He sits on the table the way a mother bird sits on her nest. His left leg has gone off into the same uncomfortable-looking angle it was at a week ago. We thought it was fully healed. Now it looks like a bit of wire hanging off his body. It does not look like a leg. It is worrying. When he stands on the dish taking sips of water, he can hardly stop himself from falling into the water. Balancing on one leg is no treat. Luckily flying is no problem for him. I use a piece of Kilkenny limestone to gently smash his biscuits into small pieces. He does not fly away when I mash. He stays close by waiting until I stop so that he can begin eating.

31 August

Michael was being badly bullied by some bigger fatter robins this morning. He walked into the kitchen so I fed him some cracker crumbs on the floor. He ate in his sitting-down position in peace and quiet.

3 September

Michael now sits on my knee. He sits on Simon's shoulder. He has not sat upon my shoulder and he has not sat upon Simon's knee. Today he sat on Maud's foot. He is completely happy to be on us and near to us. Sudden movements frighten him but mostly he appears to like the sound of voices and the presence of people. He made a diving attack on some other robins who showed up and started to eat some of his crumbs so I am less worried about him being able to survive than I was. His leg is bent but it does not stop him from flying. It does not dangle from his body in such a useless way.

8 September

The winds are wild and gusty. We are being buffeted about. The sound of the wind is never not in my ears. It is always in my ears. I hear it when I am inside the house and I hear it when I am outside the house. I hear it while I sleep and while I eat. I hear it while I am thinking of other things. I am worried about Michael. I have not seen him all day. I hope he is tucked away somewhere safe and out of the wind. I think he spends a lot of time under the large rosemary bush. I hope he has plenty to eat.

11 September

The wind never stops. It never stops. It is exciting and it is completely annoying. I cannot remember how long it has been. It seems like it has been windy forever. I feel we could be blown away. We might end up in another country or at least another county. The sun has come out and in between glorious bright sunshine there are small amounts of rain. The rain falls while the sun shines. Every few minutes the day is different. Some robins have appeared around the table. They are here to eat crumbs but none of the robins are Michael. I am looking carefully at their markings and at their legs, which are all strong and straight. I am looking for the crooked tail feather. I am worried. I thought maybe all the robins had gone away but that is not the case. I wonder if Michael has been chased away by bigger birds or if he has been blown away by the wind or if he has been the victim of a bigger creature. Some of the robins are here. Where is Michael?

13 September

I am hoping that Michael has found a new place to call his own. I do not know if the other robins chased him away. There are two robins who still stop at the table regularly. They both have fat bodies and strong straight legs. They are not Michael. I was worried about him in the wild winds but the winds have stopped now. I thought he might be sheltering but if that was the case he should be back. I have spent time in the places where he used to go with me. I have picked raspberries and talked to him as if he was nearby. I hoped that if he was hiding my voice might encourage him to come out. I have done some weeding. I have sat on the kitchen bench and I have drunk tea out at the big table always hoping and hoping he might appear. I have been hoping his curiosity would make him come along to see what I was doing. One day I sat on the bench in the rain under an umbrella just in case he felt the rain provided a safe time to come out of hiding. Most birds are not out in the rain. I hope wherever he is that he is happy. He might not remember our voices and our treats. The brain of a robin might not hold on to a past. Maybe the present is enough. I still look outside for him many times each day. I hope he is not dead.

NOTES ON CONTRIBUTORS

LATIFA AKAY is a writer from Belfast. Her prose and poems have appeared in places including *The Good Journal*, *Popshot* magazine, and *Poetry Birmingham Literary Journal*. She lives and works in London and formerly worked as a journalist in Istanbul.

SARA BAUME is the author of three novels, the most recent of which is *seven steeples*, and one book of non-fiction, *handiwork*. She is based in West Cork where she also works as a visual artist.

JOHN BERGER (1926–2017) was a British essayist and cultural thinker as well as a prolific novelist, poet, translator, and screenwriter. He is best known for his novel *G.* and his book and BBC series *Ways of Seeing*.

JUNE CALDWELL is the author of the short story collection *Room Little Darker* and her debut novel, *Little Town Moone*, is forthcoming from John Murray.

NIAMH CAMPBELL won the *Sunday Times* Audible Short Story Award in 2020 and the Rooney Prize for Irish Literature in 2021. Her novels *This Happy* (2020) and *We Were Young* (2021) are published by Weidenfeld & Nicolson. She lecturers in creative writing at UCD.

VONA GROARKE's latest, eighth, poetry collection is *Link: Poet and World* (Gallery Press, 2021). *Hereafter*, a formally innovative study of Irish women domestic servants in 1890s New York, was published by New York University Press in 2022.

EDWARD HOAGLAND is an American novelist, travel writer, and essayist, noted especially for his writings about nature and wildlife. His novels and collections of essays include, among many others, *Cat Man* (1956), *The Peacock's Tail* (1965), *The Moose on the Wall* (1974), *Red Wolves and Black Bears* (1976), *Hoagland on Nature* (2003), and *In the Country of the Blind* (2016).

TIM MACGABHANN's first two novels, *Call Him Mine* and *How to Be Nowhere*, appear with Weidenfeld & Nicolson. He lives in Mexico.

SABRINA MANDANICI is an art critic and writer, based in Berlin and New York City. Currently, she is a PhD candidate in Creative Writing at Bath Spa University, researching art criticism as a form of storytelling.

DARRAGH McCAUSLAND is a writer from Kells, Co. Meath. He has had fiction and essays published in *The Dublin Review*, *Lighthouse*, *Stonecutter*, *gorse*, *The Tangerine*, and *The Pig's Back*.

HONOR MOORE is an American writer of poetry, creative non-fiction, and plays. Her latest book, *Our Revolution: A Mother and Daughter at Midcentury*, was published by W. W. Norton in 2022. Her work has appeared in *The New Yorker*, *The Paris Review*, *The*

NOTES ON CONTRIBUTORS

American Scholar, and many other journals and anthologies. She lives and writes in New York, where she is on the graduate writing faculty of The New School.

EILEEN MYLES (they/them) is a poet, novelist, and art journalist whose practice of vernacular first-person writing has become a touchstone for a cross-generational readership. *Pathetic Literature* is their latest. 'Millie' is from *All My Loves*, a work in progress.

STEPHEN SEXTON is the author of two books of poems: *If All the World and Love Were Young* (2019) and *Cheryl's Destinies* (2021), published by Penguin.

JESSICA TRAYNOR is a poet, essayist, librettist, and poetry editor at *Banshee*. Essays have appeared in *Tolka*, *Banshee*, *The Dublin Review*, and *Winter Papers*, and have been longlisted for the Deborah Rogers Foundation Award and the Fitzcarraldo Editions Prize. *Pit Lullabies* (Bloodaxe, 2022) is a Poetry Book Society Recommendation.

ERICA VAN HORN is an American artist and writer who lives in Tipperary where, together with Simon Cutts, she works on the publications and projects of Coracle; somewordsforlivinglocally.com, her online journal, has been the source of several books.

SUZANNE WALSH is an artist and writer working with performance, audio, and text. They have an interest in non-human worlds and in creating rifts through which new meanings and realities can emerge. They publish essays, art writing, poetry, and fiction in publications including *gorse*, *Fallow Media*, and *Winter Papers*.